HILLCROFT COLLEGE

00020589

155.2
COR

D0297748

Consciousness and Human Identity

Consciousness and Human Identity

Edited by

JOHN CORNWELL

Jesus College
Cambridge

HILLCROFT COLLEGE
LEARNING RESOURCES
SOUTH BANK
SURBITON
SURREY KT6 6DF

OXFORD   NEW YORK   TOKYO
OXFORD UNIVERSITY PRESS
1998

Oxford University Press, Great Clarendon Street, Oxford OX2 6DP

Oxford New York
Athens Auckland Bangkok Bogota Bombay
Buenos Aires Calcutta Cape Town Dar es Salaam
Delhi Florence Hong Kong Istanbul Karachi
Kuala Lumpur Madras Madrid Melbourne
Mexico City Nairobi Paris Singapore
Taipei Tokyo Toronto Warsaw
and associated companies in
Berlin Ibadan

Oxford is a trade mark of Oxford University Press

Published in the United States
by Oxford University Press Inc., New York

© *The contributors listed on p. xvii, 1998*

All rights reserved. No part of this publication may be
reproduced, stored in a retrieval system, or transmitted, in any
form or by any means, without the prior permission in writing of Oxford
University Press. Within the UK, exceptions are allowed in respect of any
fair dealing for the purpose of research or private study, or criticism or
review, as permitted under the Copyright, Designs and Patents Act, 1988, or
in the case of reprographic reproduction in accordance with the terms of
licences issued by the Copyright Licensing Agency. Enquiries concerning
reproduction outside those terms and in other countries should be sent to
the Rights Department, Oxford University Press, at the address above.

This book is sold subject to the condition that it shall not,
by way of trade or otherwise, be lent, re-sold, hired out, or otherwise
circulated without the publisher's prior consent in any form of binding
or cover other than that in which it is published and without a similar
condition including this condition being imposed
on the subsequent purchaser.

A catalogue record for this book is available from the British Library

Library of Congress Cataloging in Publication Data
Consciousness and human identity / edited by John Cornwell.
Includes index.
1. Consciousness–Physiological aspects. 2. Cognitive
neuroscience. 3. Identity (Psychology) I. Cornwell, John.
QP411.C576 1998 612.8'2–dc21 98-18360
ISBN 0 19 850323 7 (Hbk)

Typeset by Downdell, Oxford
Printed in Great Britain by
Biddles Ltd
Guildford & King's Lynn

Preface

The phenomenon of human higher-order consciousness—that there is something that it is like for human beings to experience the world—has puzzled philosophers, naturalists, and theologians down the ages. Now, and belatedly, consciousness has caught the interest of contemporary scientists, some of whom believe they are on the brink of discovering its basis in neurobiological processes. This book, composed of contributions from scientists and philosophers, is about the prospects for finding a scientific explanation of consciousness. It is also about how a scientific explanation could affect our view of ourselves.

What processes in the brain, the central nervous system, asks the neuroscientist, account for the sense we have of looking out at the world from the secret theatre of the self? What explains that uniquely personal experience we have of smelling a rose, or feeling the pain of toothache, agonizing to me and to nobody else, or seeing the point of a *New Yorker* cartoon, or sensing a pang of post-modernist angst in the run up to the millennium? Moral and social behaviour—especially, a sense of worthy and unworthy actions, artistic expression and aesthetic appreciation—involve an account of human nature that crucially includes conscious awareness, conscious autonomy, both in oneself and in others. The importance of discovering the link between our physical brain states and consciousness needs no special pleading, for consciousness lies at the very heart of what it means to be human: in the Western tradition consciousness is the very eye of the human soul.

Scientists' neglect of consciousness through much of the twentieth century seems at first sight strange. The neuroscientist Eric Kandel wrote grandiloquently in the preface to his 1985 edition of *Principles of neuroscience* that 'one of the last frontiers of science, perhaps its ultimate challenge, is to understand the biological basis of mentation'. But why, at that time of writing, had neuroscientists been so laggard in a century marked by huge strides in most other scientific explorations of nature? As neuroscientists put it, the vulnerability and vast complexity of the brain kept researchers at bay. They did not have the tools or techniques to enter the living cortex without devastating what they explored. This has not daunted the ambitions of those cognitive scientists who believe that real brains are as dispensable to the study of thinking as feathers are to aerodynamics; hence some researchers in the field of artificial intelligence were optimistic about replicating consciousness with silicon and circuitry. Some even believed that the day would come when humans would down-load their minds into suitable software and so initiate an immortal existence.

By the late 1980s, however, the prospects for neuroscience were transformed. Just as the invention of the telescope and the discovery of mathematical physics gave rise to new ways of understanding the Universe, so rapid advances in non-invasive brain imaging, and techniques of measuring at a micro level the activity of individual cells, receptor sites, and neurotransmitters, revolutionized the exploration of the brain and the central nervous system. At the same time, the increased power of computers enabled researchers to test their theories with ever more sophisticated models. After a century in the doldrums, neuroscience was, in its own estimation, on a voyage to the final frontiers of science.

But the idea that the 1990s would be special to brain research was due not solely to the inspiration of academic science. A crucial impulse came from the burgeoning biotech industry and the expanding neuroscience arm of the pharmaceutical industry whose marketing departments announced a new age of rationally designed brain drugs with cure-alls for everything from pain to Parkinson's disease. Their clinical allies in genetics and neurology, eager to explore revolutionary techniques in grafting, implanting, genetic screening, and 'carrier' therapies to alter 'defective' genes, were not slow to advance the cause. On 1 January 1990, the lobbyists were rewarded with a joint resolution of the House and the Senate of the United States to designate the nineties 'The Decade of the Brain'. A principal stated reason behind the initiative was the official estimate that some $350 billion was being lost to the US economy each year through brain-related ills, including depression, Alzheimer's, and the consequences of aggressive behaviour (which alone attracted grants of $500 million for neurogenetic research in 1994).

This promise of major social and medical amelioration was driving the momentum and direction of investment and funding in neuroscience. But even as the discipline was being feted in anticipation of a host of social and medical benefits, there were growing expectations, evident in proliferating symposia and published academic papers, that neuroscience had a significant role to play in the Holy Grail of Kandel's 'basis of mentation'. Neuroscience, proclaimed a new constituency of philosophers, psychologists, and philosophically interested neuroscientists, would provide an unprecedented approach to explaining the link between our physical brain states and consciousness.

The first international symposium on the science of consciousness was announced for April 1994 at the University of Arizona at Tucson. A thousand delegates participated, including many leading scientists, mathematicians, and philosophers. There were 55 papers covering themes as disparate as 'quantum computation in the neural membrane' and 'consciousness and blind-sight', 'self-awareness in Alzheimer's patients', and the application of 'positron emission tomography to emotion'. In a ceremony that concluded the meeting these representatives from a host of academic backgrounds literally joined hands. There had indeed been curious alliances between themes as diverse as quantum physics and anaesthesiology, commissurotomy and pharmacology, and the meeting prompted a remarkable proliferation of international research programmes, seminars, and papers. When the first *Journal of Consciousness Studies* was launched in 1994 it published a bibliography of more than 1000 specialist articles.

Meanwhile a new genre of consciousness book had been making its appearance in publishers' lists, to mention just a few: Roger Penrose's *The emperor's new mind* (1989), Daniel Dennett's *Consciousness explained* (1991, John Searle's *The rediscovery of the mind* 1992, Gerald Edelman's *Bright air brilliant fire* 1992, and Francis Crick's *The astonishing hypothesis* (1994). It was clear, from the torrent of papers and books, that the scientific quest for consciousness was nothing if not multidisciplinary. It was also clear that scientists and philosophers of mind, despite the elucidations of neuroscience, were not about to agree on anything like a single theory.

There were profound disagreements, to begin with, over the definition of consciousness itself, leading to fundamental clashes over the prospects for enlightenment. The British philosopher Colin McGinn, for example, argued in his *The problem of consciousness* (1991) that consciousness was resistant to explanation in principle and for all time; while the American philosopher of mind Daniel Dennett—a prominent figure at the Tucson conference—countered that a complete explanation not only lay ready to hand, but would soon be exemplified by the construction of a conscious machine called COG. The nub of the divide, however, was the issue of methodology itself—the prospects for scrutinizing consciousness by means of reductionist science. Could this well-tried procedure, by which scientists come to understand phenomena by examining their most reducible physical aspects and processes, solve anything so elusively holistic and phenomenological as the mind–brain relationship?

Students of consciousness inclined to sympathize with this scruple were by 1994 in the habit of talking about the 'hard problem' and the 'easy problem' of consciousness, in order to highlight a central paradox. While acknowledging that reductionist science would make progress in solving the so-called 'easy' problems—objective accounts, for example, of the neurophysiology of vision—the 'hard problem' of consciousness, the phenomenological problem, would persist. David Chalmers, in the opening discussion of the Tucson conference, described it like this:

> It is undeniable that some organisms are subjects of experience. But the question of how it is that these systems are subjects of experience is perplexing. Why is it that when our cognitive systems engage in visual and auditory information processing, we have visual or auditory experience: the quality of deep blue, the sensation of middle C? How can we explain why there is something it is like to entertain a mental image, or to experience an emotion?

There was nothing particularly new about the separation of the 'hard' and the 'easy' problems of consciousness; philosophers like Thomas Nagel and John Searle had done much to popularize the related problem of 'qualia' since the 1970s, warning against easy solutions, or definitions, that fail to get to grips with the central riddle. What seemed new was the vehemence of the confrontation between those who recognized the authenticity of the 'hard' problem and those who seemed to think that it was illusory, destined to fade in the light of reductionist knowledge about the brain. Also new were indications that dualism—the belief that there are two fundamentally different kinds of phenomena in the world, minds and bodies—was making a fitful comeback. The Oxford mathematician Roger Penrose's attachment to the idea that consciousness can be explained by a special sort of non-classical physics in the brain appeared to support such a contention, borne out by the curious circumstance that his closest ally in the consciousness debate, the late Sir John Eccles, was a distinguished neuroscientist and fierce expositor of substance dualism. The philosopher David Chalmers, moreover, was giving notice of what he called a *psychophysical* solution, a non-reductive theory of consciousness that would stem from basic laws of nature as yet unknown. 'This position', Chalmers was conceding, 'qualifies as a variety of dualism, as it postulates basic properties over and above the properties invoked by physics.'[1] Issues of human dignity and autonomy, of course, were never entirely absent in such debates; fuelling the growing confrontation between the reductionists and the anti-reductionists was an urgent reformulation of an old question: are we, or are we not, closer to the machines of our own devising than we should care to admit?

Conscious of a sense of crisis about reductionist methods in a variety of areas of the natural sciences, the Science and Human Dimension Project at Jesus College, Cambridge, had convened a symposium on that topic in the autumn of 1992. Had reductionism—so crucial to the success of modern science and technology—reached its limits? Or was reductionist science merely surrendering to a new wave of relativism that insists on an equivalence between science and any other mode of enquiry?

At that meeting, Freeman Dyson, while expressing reasonable sympathy with an easy equivalence between the realms of science and of art (science, he said, 'is an alliance of free spirits rebelling against the local culture'), praised reductionism's attempt 'to reduce the world of physical phenomena to a finite set of fundamental equations'. He then went on to cite the work of Schrödinger and Dirac in 1925 and 1927 on quantum mechanics as great intellectual victories of the century—'bewildering complexities of chemistry and physics ... reduced to two lines of algebraic symbols'.[2] Yet despite such triumphant citations the meeting quickly divided into two camps—on the one hand, the unashamed reductionist radicals, exemplified by Peter Atkins's bold 'The limitless power of science', and, on the other, the moderates who sought to combat what Mary Midgely called 'Reductive megalomania'.

Later, in a review of the published papers of that symposium in *Nature*,[3] John Casti described these symposium moderates—for example, Midgely, Penrose, and the neuroscientist Gerald Edelman—as the 'good guys in white hats', in contrast to the 'bad guys', who included neurophilosopher Patricia Churchland, the chemist Peter Atkins, and the astronomer John Barrow. Casti is a collaborator at the Sante Fe Institute, where an interest in complexity theory has encouraged an emphasis on understanding nature from dynamic, emergent, and relational perspectives. His review eloquently emphasized nature's complex combinations of structure and openness, law and chance, order and chaos, determinism and probability, and even suggested that anti-reductionism had the monopoly of some sort of moral high ground. Less complimentary to the 'white hats', however, was the physicist Steven Weinberg, who argued in another valuable review of the proceedings,[4] that the participants had surely failed to make a clear distinction between reductionism as a programme for scientific research, which he endorsed, and reductionism as a view of nature, about which he was not so sure. 'For instance,' he wrote,

> the reductionist view emphasizes that the weather behaves the way it does because of the general principles of aerodynamics, radiation flow, and so on (as well as historical accidents like the size and orbit of the earth), but in order to predict the weather tomorrow it may be more useful to think about cold fronts or thunderstorms. Reductionism may or may not be a good guide for a program of weather forecasting, but it provides the necessary insight that there are no autonomous laws of weather that are logically independent of the principles of physics.

Weinberg's reasonable critique seemed to give the radicals the benefit of the doubt. And yet, there was one notable and outstanding problem: the project of subjecting consciousness to reductionist investigation. With Edelman, Penrose, Churchland, Margaret Boden, and the late Hao Wang on the list of speakers, it was not surprising that the issue of consciousness had haunted the sessions and lingered in the corridors afterwards. While the radicals vigorously defended their corner on most other issues, the majority of participants appeared to concur with Gerald Edelman's spirited conclusion that the attempt to 'put mind back into nature' constitutes the end of the old Enlightenment and the beginning of the new. The 'neuroscience revolution', he maintained, was as historically significant as the Copernican revolution in cosmology—with crucial bearings on reductionism. Despite the success of reductionism in physics, chemistry, and molecular biology, 'it becomes silly reductionism when it is applied exclusively to the matter of the mind'. The workings of the mind, he went on, 'go beyond the description of temporal succession in physics. Finally, individual selfhood in society is to some extent a historical accident.'

This remarkable reflection was echoed by Weinberg himself with the admission that 'consciousness poses the greatest challenge to reductionism ... It is difficult to see how the ordinary methods of science can be applied to consciousness, because it is the one thing we know about directly, not through the senses.' Having praised Peter Atkins for his 'in-your-face reductionist polemic', Weinberg finally dissociated himself from him for not being 'sufficiently sensitive to the problems surrounding consciousness'.

It was in order to explore some of those 'problems surrounding consciousness' and the application of 'ordinary methods of science' to consciousness, that the Science and Human Dimension Project convened a second conference in the autumn of 1995. Benefiting from the contributions of a great many works of popular and academic exposition on the subject, as well as the newly founded *Journal of Consciousness* and the first Tucson symposium, we were keen to take the debate forward beyond the arguments about reductionism.

Clearly the consciousness debate contains all manner of loaded questions and perspectives, and every declaration of viewpoint is coloured by background philosophy of mind and philosophy of science. The Science and Human Dimension Project had endorsed a moderate standpoint on the issue of reductionism: it was guardedly optimistic, therefore, about finding the neurobiological causes of consciousness, and yet it did not underestimate the philosophical problems. The project had declared itself philosophically in 1992, to the extent of steering clear of substance dualism, and it recognized the danger of lapsing into other forms of dualism by proxy—the tendency to see the mind–brain relationship as similar to hardware and software in a computer process. We did not warm, moreover, to functionalism—the notion that there was nothing special about neurobiology, about evolutionary biology, in the development of consciousness.

Our aim, then, was to press forward and to present and discuss a restricted repertoire of scientific studies of consciousness, while broadly accepting Steven Weinberg's scepticism about exclusively reductionist approaches to understanding nature. We wanted to hear from researchers in neurobiology of consciousness who had worked both with computer models of biological processes as well as with animal models. Having been impressed with Gerald Edelman's report on his theory of neuronal group selection in 1992, we were keen to hear again from his newly founded Neurosciences Institute in La Jolla, California. Accordingly, we invited Edelman's colleague, Olaf Sporns, to give a paper on the modelling of a 'selectionist'-based system and the role of 'populations' and 'variability' in the latest phase of the institute's 'Darwin Machine' programme. It is Edelman's view that while his neural Darwinism approach to the mind–brain relationship is unlikely to produce a complete explanation for consciousness, it is a feasible candidate for elucidation.

In contrast, we asked Bernard Balleine and Anthony Dickinson, psychologists with a background in neurophysiology, to report on laboratory work with animals in which they had been exploring the interface between affect and cognition, leading to a study of motivational control of goal-directed action. Balleine and Dickinson are convinced that the future of consciousness studies depends on a deeper understanding of affective processes and how desires arise from the physiological systems that ultimately determine our evaluation of the goals of our actions.

Since our project's work aims to point back continually at the human dimension of science, we were fortunate to be joined by another biologist, Steven Rose, whose paper sets mind–brain explorations in the context of sociobiology and the politics and economics of the 'Decade of the Brain'.

William Clocksin, of the Computer Science department at Cambridge, had contributed a valuable paper on artificial intelligence (AI) and the issue of representationalism at the 1992 symposium. He joined us again to speak on the theme of how artificial intelligence research is influenced by 'underlying tacit assumptions concerning human identity'. His critical approach involves challenging implicit normative 'loading' of AI and calls for adjustment if progress is to be made in the field. His plea that future research should look at the idea of 'narrative architectures' built on a foundation attentive to the way signals and symbols influence the way we make sense of experience, is both original and intriguing.

Another controversial aspect of non-biological approaches to consciousness is that quantum physics might play a crucial role in understanding how brain processes give rise to consciousness. Quantum physics, moreover, has prompted fresh ways of understanding the relationship between mind, matter, and physicalism. In 1992 Roger Penrose had briefly outlined his thesis that microtubules in the cytoskeletal structures of neurones are a potential site for quantum physical activity; there had also been discursive exchanges on the implications of Kurt Gödel's theorem and the non-algorithmic nature of mental processes. We now invited Jeremy Butterfield, a Cambridge philosopher of science with a special interest in quantum physics, to provide a critical discussion on the theme of quantum theory and the mind against the background of a rapidly growing literature on that subject. Quantum theory and its implications for philosophy is an extremely difficult area and there is no painless way through various stages of the debate. Butterfield provides a guide to how quantum physics gives rise to new ways of understanding 'physicalism' and the mind–matter divide.

The symposium was buttressed by philosophical scene-setting and analysis. Margaret Boden provides a survey and recent history of the principal contentions in philosophy of mind, with critical discussion of the scepticism of Jerry Fodor and Colin McGinn on the possibility of an explanation for consciousness.

John Searle, a veteran in the field of consciousness studies, comes down clearly on the side of those who advocate a scientific quest for consciousness. His contribution distinguishes between two versions of the subjective–objective divide: an epistemic version and an ontological version. This distinction, he argues, reveals a fallacy in the pessimism about studying consciousness scientifically. Science, goes the fallacy, is objective, but consciousness is subjective, hence science cannot explore consciousness. The source of the fallacious syllogism, argues Searle, is an ambiguity between the two versions of the subjective–objective divide. Science is objective epistemically since its conclusions are intersubjective agreements about the facts. Consciousness is ontologically subjective, for consciousness appears to depend on personal experience. But ontological subjectivity does not entail epistemic subjectivity. Hence it is possible to have an epistemically objective, scientific investigation into the nature of ontologically subjective consciousness!

This was a symposium in which the speakers were encouraged to reflect, in the final analysis, on the implications of a scientific understanding of consciousness for traditional views of human nature. Mary Midgley's contribution, 'Putting ourselves together again' ponders the category mistakes commonly made in discussion about consciousness. It was inevitable that she should raise the question of 'soul' language, a term associated with consciousness down the centuries. As a token of the historic importance of consciousness in the Western tradition of understanding the human soul, we completed the symposium by inviting reflections from a philosopher of science and religion, and also from a theologian.

Fraser Watts inhabits a peculiarly privileged vantage point in the consciousness debate as the first Starbridge Lecturer in Theology and Natural Science at Cambridge University. Since the seventeenth century notions of consciousness have been clouded by a Cartesian view of the body–soul divide, widely and mistakenly associated to this day with a religious view of consciousness and the human person. Watts shows not only how religion has moved on from this position but reminds us of the tension that has existed between unitary and dualistic views of the soul ever since Plato and Aristotle. The same point is forcibly made by Nicholas Lash, who points out in what he calls a series of 'grace notes' on John Searle's *Rediscovering the mind* (1993) that as a theologian with 'quite conventional views on Christian doctrine', he welcomes Searle's 'biological naturalism' as wholeheartedly as he welcomes 'Gerald Edelman's attempts to put the mind back into nature'. Lash claims that popular substance–dualist construals of the relationship between the human body and its life, or soul, have for centuries been an embarrassment to mainstream non-dualist Aristotelianism (to say nothing of being incompatible with the Jewish anthropology that the Christian Scriptures breathe), and hence questions the extent to which 'Cartesian dualism is a recent disruptive *innovation* in the history of Christian anthropology'. He playfully expresses a hope that some of his remarks might 'flush out the residual Cartesianism in some of my scientific friends who suppose themselves quite free from the disease!'

The final word is left to Peter Lipton, who reviews and analyses the drift of the symposium in terms of kinds of explanation. 'If we aim to explain consciousness,' he asks, 'which sort of explanation should we seek?' Against the background of so many contributions in which terms such as 'explanation', 'causes', and 'accounts' have been freely used, both by scientists and by philosophers, Lipton's discussion about the distinction between identity explanations and causal explanations is salutary and crucial. His endorsement of the project—how consciousness should be studied scientifically—is, moreover, enthusiastic. It is a mistake, he declares, for scientists to take the view that consciousness is not a fit subject for science. 'Scientists should study consciousness, but they should seek causal explanations, not identity explanations. For the difficulties disappear or are at least substantially diminished if we shift our attention from the question of what mental states *are* to questions of their aetiology.'

The chapters that form this book are based on papers read at the Jesus College symposium. Most have been reworked over a period of two years in the light of our discussions and further developments.

The symposium was made possible by the generosity of the Mrs L. D. Rope Third Charitable Settlement and the encouragement of Mr Crispin Rope. Thanks are also due to Jeremy Butterfield, Peter Lipton, Simon Kidd, and Jonathan S. Cornwell. Finally I must thank the Master and Fellows of Jesus College, Cambridge, for their enthusiastic support.

Cambridge   J.C.
March 1998

References

1. Chalmers, D. (1996). Facing up tot he problem of consciousness. In *Towards a science of consciousness: the first Tucson discussions and debates* (ed. Stuart R. Hameroff, Alfred W. Caszniak, and Alwyn C. Scott), p. 17. MIT Press, Cambridge, Mass.
2. Dyson, F. (1995). The scientist as rebel'. *The New York Review*, 25 May, 31–3.
3. Casti, J. (1995). Review of *Nature's imagination*, ed. John Cornwell, (Oxford University Press, Oxford, 1994). *Nature*, **374**, 840.
4. Weinberg, S. (1995). Reductionism redux. *The New York Review*, 25 October, 29–42.

Contents

Contributors

Bernard Balleine
Assistant Professor in the Department of Psychology, University of California, Los Angeles

Margaret A. Boden
Professor of Philosophy and Psychology in the School for Cognitive Sciences at the University of Sussex. Author of *The creative mind* and *Artificial intelligence and natural man*

Jeremy Butterfield
Senior Research Fellow, All Souls College, University of Oxford

W. F. Clocksin
Lecturer in Computer Science at the University of Cambridge, and Fellow of Trinity Hall. Author (with C. S. Mellish) of *Programming in PROLOG*

John Cornwell
Director of the Science and Human Dimension Project and Fellow of Jesus College at the University of Cambridge. Editor of *Nature's imagination*, and author of *Power to harm*

Anthony Dickinson
Lecturer in the Department of Experimental Psychology at the University of Cambridge

Nicholas Lash
Norris Hulse Professor in the Faculty of Divinity at the University of Cambridge. Author of *Easter in ordinary* and *Theology on Dover Beach*

Peter Lipton
Professor of the History and Philosophy of Science at the University of Cambridge. Author of *Inference to the Best Explanation*

Mary Midgley
Formerly Lecturer in Philosophy in the University of Newcastle upon Tyne. Author of *Wisdom, information and wonder* and *Science as salvation*

Steven Rose
Professor of Biology and Director, Brain and Behaviour Research Group at the Open University. Author of *The making of memory* and *Lifelines* and editor of *From brains to consciousness?*

John R. Searle
Mills Professor of Philosophy at the University of California, Berkeley. Author of *Rediscovering the mind* and *The construction of social reality*

Olaf Sporns
The Neurosciences Institute at San Diego, California

Fraser Watts
Starbridge Lecturer in Theology and Natural Science in the Faculty of Divinity, and Fellow of Queen's College at the University of Cambridge

···

CHAPTER 1

Consciousness and human identity: an interdisciplinary perspective

MARGARET A. BODEN

1 Introduction

Many scientists, and also some philosophers, predict that the next century or two will be the age of neuroscience—in which the mysteries of consciousness will finally be revealed. Is this post-millennial scientific triumph likely?

Jerry Fodor thinks not (Fodor 1992). I hesitate, for obvious reasons, to describe Fodor as the Mike Tyson of philosophy. But you have to admit that he always comes out fighting: he does not give up easily. Yet this is his verdict on the possibility of a 'science of mind':

> Nobody has the slightest idea how anything material could be conscious. Nobody even knows what it would be like to have the slightest idea how anything material could be conscious. So much for the philosophy of consciousness.

Of the many philosophical problems regarding consciousness, the foremost is how to explain, or even to acknowledge, first-person subjective experience from the objective viewpoint of science. Thomas Nagel (1974), for instance, claims that we could know everything there is to know about the objective facts of bat behaviour and neurophysiology, without knowing 'what it is like' to be a bat. Frank Jackson (1982) argues, similarly, that someone confined to a black-and-white scientific laboratory, with no windows and only black-and-white books on the library shelves, could in principle master the science of colour vision (all of the relevant physics, biology, and experimental psychology). But they would not know what it is like to see colour: this, they could learn only by being allowed to look out of a door or window onto the outside world. And anti-functionalists generally complain that functional-ism—widely popular with physicialists since its inception at mid-twentieth century (Putnam 1960, 1967)—cannot account for qualia, for first-person phenomenal experience.

Functionalism defines mental states in terms of their causal relations between perceptual input, motor output, and other mental states (abstractly defined, likewise). It has many attractions for philosophers sympathetic to science, as compared with its scientifically inclined predecessors: epiphenomenalism (Huxley 1893), behaviourism (Watson 1913), and the mind–body identity theory (Place 1956; Smart 1959). For it saves the existence, and the causal efficacy, of internal mental states. It saves their conceptual connection with behaviour, and with other mental states. It posits a physicalist base for each individual token of a mental state, without arguing that the physiology of pain, hope, or fear must be the same in all species. It allows that the categories of folk psychology may be scientifically useful. And it promises to define the abstract functional relations concerned in a clear and testable way, by making use of computational concepts.

Many scientifically inclined philosophers welcomed functionalism with open arms—but not, their opponents would say, with open minds. Some anti-functionalists object that functionalism pays too little regard to science (more specifically, to neuroscience), because the actual biological base of mentality in a given individual or species is seen by functionalism as philosophically irrelevant (Searle 1992). The anti-functionalists' main complaint, however, is that functionalism cannot explain qualia, nor even admit their existence. Obviously, they say, one can imagine all the abstract functions being fulfilled by some system, such as some robot, without there being any qualitative experience involved. (This use of 'obviously' is questionable: see section 3.)

If qualia and subjectivity comprise the foremost puzzle of consciousness, personal identity is close behind. It seems to have some very intimate—indeed, essential—connection with consciousness. But just what is it, and how is it to be explained? Perhaps Fodor would be prepared to repeat his pessimistic conclusions in respect of a science of human identity. No one, he would then say, knows what it would be like to have the slightest idea how anything material could have, or be, a personal identity.

If Fodor is right, this chapter—and this entire volume—is a waste of time. (A knock-out blow, indeed.) Up to a point, I share his pessimism. But I think that something, though admittedly not much, can be said about what it would be like to have a science of consciousness. And that 'something' shows why various neuroscientific discoveries recently hailed by enthusiasts as solutions to problems of consciousness are not genuine solutions. At best, they provide empirical information that might eventually contribute to such solutions. At worst, they are utterly irrelevant to any philosophically fundamental account of consciousness, being founded on an epistemological distinction that only a metaphysically prior human subject can make. Whether we choose the 'best' or the 'worst' interpretation depends on our acceptance of Cartesian or non-Cartesian presuppositions, respectively.

2. Correlations of consciousness

Neuroscience has made enormous advances since the mid-twentieth century. The anatomy, physiology, biochemistry, and computational functions of the brain are hugely better understood than when I studied these matters as a medical student, 40 years ago. These discoveries were made possible by ingenious experimental techniques such as single-cell recording, radio-active marking, and the new methodologies of brain-imaging. The myriad experimental data provide growing evidence of detailed correlations between neural structures and processes on the one hand and conscious phenomenology on the other (Milner and Rugg 1992; Posner and Raichle 1994).

To give just one example, recent work suggests that a distinct group of cortical neurones is active only when a particular type of visual phenomenology is being experienced (Logothetis and Schall 1989). If a pattern of vertical bars is continuously presented to one eye, and a pattern of horizontal bars to the other, what a human subject actually sees is a regular alternation between the two (simultaneously presented) patterns. If a monkey is put in the same experimental situation, its discriminatory behaviour alternates in a way that suggests that it, too, sees first the one pattern and then the other. Moreover, a group of cells in the monkey's visual cortex is active only when its behaviour suggests that it is *aware* of the vertical pattern; a different group is continuously active while the vertical pattern is being presented as a stimulus to the eye. The same applies in respect of the horizontal patterns. This experiment is even more intriguing than those (many) that show merely that the presentation of a particular physical stimulus is reliably correlated with activity in particular neurones.

Apparently, then, we have come a long way since Descartes. But have we, really? Descartes would be just as intrigued as we are by the specific neuroscientific discoveries that have been made. He would probably be taken aback by the 'alternating perceptions' experiment just described, which seems to discriminate between conscious and unconscious responses in a non-human animal. And he might well be astounded by the 588-page volume on the pineal gland recently published by the New York Academy of Sciences, which describes its functions as a pacemaker in ageing and carcinogenesis. But he would not be at all surprised that some such discoveries have occurred. Indeed, he predicted them.

He argued that brains are complex material systems amenable to scientific investigation like any other, and that detailed mind–brain correlations exist— if only in human beings. Despite doing very few experiments (and those clumsily performed, even by the standards of his time), Descartes is honoured as the father of experimental physiology, including not only neurophysiology but psychophysiology too. In short, he posited correlations between brain and consciousness, and he encouraged scientists to look for them.

Descartes recognized various philosophical difficulties associated with this scientific project. One was the problem of how Cartesian mind could affect brain-matter, which he described to Princess Elizabeth of Bohemia as 'the question people have most right to ask me in view of my published works'. In addition, his initial suggestion that the pineal gland (at the behest of the soul) impels the animal spirits within the brain conflicted with the principle of the conservation of motion. So he suggested instead that the gland introduces no new motion into the brain, but—by moving on its slender stalk 'like the rudder of a ship'—merely guides the cerebral fluid in one direction rather than another. But even this analogy is problematic. The stalk's function as a rudder can be understood only in terms of spatial reflection, yet consciousness (according to Descartes) is not spatial.

No one nowadays believes the pineal gland has any privileged role in relating mind to body. But contradictions comparable to that just noted still appear.

Consider Francis Crick's claim, that 'freewill is located in or near the anterior cingulate sulcus' (Crick 1994, p. 268). Whether we take 'freewill' to connote a conscious act of choice, or a control mechanism for scheduling potentially conflicting motives (see section 4), it is a category-mistake to describe it as being 'located' anywhere. There is no category-mistake involved, of course, in claiming that the cerebral events underlying it are located in a specific part (or distributed across a larger area) of the brain. Descartes himself would have been happy with that.

He would have been happy, too, with Crick's claim that the 'binding' of different visually discriminable properties that is necessary for the conscious perception of an object happens only when the various cell-assemblies concerned are all oscillating at the same frequency, namely 40 hertz (Crick and Koch 1990). But binding is one thing; consciousness is another. A blindsight patient, for example, can co-ordinate visual information and motor action so as to point to the position of a visual stimulus, but without that stimulus being consciously seen (Weiskrantz 1986, 1990). If the binding hypothesis is correct, then various oscillations at the relevant frequency are presumably occurring in the person's brain. But visual experience is absent. In other words, binding is an abstractly defined function, not a mode of experience.

To be sure, this distinction is questioned by those philosophers who argue that visual 'experiences' are not special qualities or events over and above the bodily processes and/or computational functions going on. Daniel Dennett (1988, 1991), for example, argues that our concepts of qualia can be cashed out in terms of discriminatory behaviour, and the computational functions required to generate it. Even if one rejects the general distinction between function and experience, however, the experimental studies of blindsight strongly suggest that the specific function of binding may occur without consciousness. If so, then to show that oscillations of a certain frequency are necessary for binding is not to prove that they are sufficient for the conscious experience of objects.

One might defend Crick from the charge of committing a category-mistake, by pointing out that some philosophers have identified brain processes with conscious events. Today's eliminative materialists (Churchland 1979, 1990), like identity-theorists before them, postulate a strict identity between conscious sensations and brain processes. On such views, experimentally discovered correlations such as those mentioned above are not correlations between events of two distinct metaphysical classes, but between two different ways of knowing about one and the same event. Ontologically, this event is best identified as it is objectively known by science, not as it is (subjectively) known by consciousness.

If eliminative materialism is correct, then Crick can indeed sensibly say that conscious events are 'located in' various parts of the brain. But this reductionist account of mind and brain is highly controversial. Even its main proponents grant that their claims are absurd on Cartesian assumptions. And they realize that (because our science and common sense are deeply imbued with such assumptions) their claims are not straightforward scientific hypotheses, to be falsified or confirmed by empirical evidence in the usual way. Rather, they are suggestions that we make a decision to change our underlying concepts of mind and body in a particular way. Only a philosophical fiat could 'establish' the identity. But, so the argument goes, advances in neuroscience will eventually make us realize that this is the most economical, and metaphysically coherent, way of speaking. I shall return to this claim in section 3. Here, let me just note that Crick writes as though his suggestion could be decided by scientific evidence alone: in truth, however, it cannot.

An even more speculative account of the material basis of consciousness has been offered by Roger Penrose (1989, 1994a,b). Being no physicist, I cannot comment in detail on his suggestions. But his claim that physics needs a fundamentally new theory ('Complete Quantum Gravity'), if true, would apply to all matter indiscriminately—to china clay just as to neuro-protein. Likewise, the microtubules that he suggests are the locus for the relevant quantum effects in the brain are present in virtually all cells. So, unless he is prepared to be remarkably generous in his ascriptions of consciousness, he needs to say a great deal more about what is special about the brain.

Recently, Penrose and Stuart Hameroff (1995) replied to critics (Grush and Churchland 1995) by saying (in part) that the microtubules in neurones are different from microtubules in other cells. They are arranged in parallel (not radially from centrioles); they are relatively stable; there are more of them, arranged in more complex networks, than in other cells; they show greater genetic variability; they transport chemical vesicles along the axon and dendritic processes; and there are neuron-specific proteins associated with them. Even if all this is so, however, the question remains why these differences give us any reason to ground consciousness in microtubules.

To answer this question, Penrose needs to show that his theory of quantum effects in microtubules is superior to theories about cell excitations and cell oscillations. In other words, he needs to show how his theory makes the brain's generation of consciousness more intelligible.

Intelligibility is a requirement that all scientific theories must meet. What one might call 'correlational' neuroscience is natural history, rather than science. As such, it is an essential preliminary to theoretical understanding. It puts empirical flesh onto Descartes's skeletal a priori claim that correlations between brain mechanisms and consciousness exist to be discovered. But it does not show why one correlation occurs, rather than another. Still less does it show why there should be any correlation at all between brain processes and consciousness. Only a powerful scientific theory could exhibit the system within the data, and the deeper explanation for it. Is Fodor right in saying that no one has the slightest idea what this might involve?

3. Possibilities of intelligibility

We have seen that mere correlation is not enough. We need intelligible correlations, so that we understand that if X happens then Y must happen—or, at least, can reasonably be expected to happen (even if this correlation has never been tested before). To add intelligibility to correlation, we need cause, structure, and isomorphism.

Suppose that future brain-imaging experiments discovered reliable correlations between a host of conscious states and distinct brain processes.

Suppose even that experimental stimulation of a certain neurone, or group of neurones, was reliably followed by conscious experience of a certain kind, suggesting that the brain–mind correlations were causal. So far, so good. Now, suppose that those conscious states that we classify together—such as seeing X and seeing Y, or seeing a dog and seeing a cat, or experiencing scarlet and crimson—were correlated with brain processes apparently located in (or distributed over) randomly scattered areas of the brain, and having randomly different measurable properties (such as oscillation frequencies). Not even the most committed physicalist would regard this as an explanation of consciousness. It would be a mere rag-bag of individually bizarre and collectively unintelligible facts.

As it is, of course, some degree of intelligible order in mind–brain correlations have already been found. Indeed, some causal correlations have been found. Among the first examples to be discovered were the mannikins (or homunculi) in the sensory and motor cortex. Excitation of certain cortical areas causes experience or movement. Mapping the experimental results onto a brain diagram, one gets two humanlike forms, topologically correct but skewed in size and shape. Hands are connected to arms, and arms to shoulders ... but the mannikin's hands and face are abnormally large, as are the thumb, index finger, and lips. That is, the most sensitive and voluntarily mobile parts of the body are connected to larger areas of sensory and motor cortex, whose activation causes experience or movement, respectively. Other topographical mappings include those found in the visual system, wherein (for example) neighbouring cells and columns of visual cortex corresponding to neighbouring positions on the retina.

But these simple topographical mappings are not enough. A greater degree of intelligibility would be available if detailed differences in the qualitative nature of conscious experience turned out to correspond to structurally isomorphic differences in neural mechanisms. In such a case, we could hope to explain why some particular experience (rather than another) happens when these particular cells (rather than others) are activated.

Consider visual orientation-detectors, for instance (Hubel 1988). The cells in striate cortex that respond to lines of varying orientations (think of the many diameters of a circle) are arranged 'around the circle' in the brain, corresponding to the geometrical position of the stimulus. Admittedly, the most systematic experiments here were done on cats and monkeys, who cannot tell us about their experiences: these must be inferred on the basis of the animal's discriminatory behaviour. But human brains contain orientation-detectors too, and human beings can be consciously aware of lines oriented in differing directions. If activity in these cells is an important contributary factor in the conscious experience of visual contours, we have here an example of a mind–body isomorphism: equivalent 'orientation-circles' in the brain and in consciousness.

Another striking example concerns the causation of motion perception (Newsome and Salzman 1993). Experiments show a remarkably systematic correlation between activity in certain orientation-selective cells in the rhesus monkey's middle-temporal visual cortex and its perceptual discriminations of movement. Before the experiment proper, the monkey is trained to move its eyes in the direction of the perceived movement. Then, stimulating a certain cell-circuit causes the monkey to behave as though it sees motion in a certain direction. Moreover, after a while these results become predictable. In other words, the relevant cells are spatially organized within the brain in a way that corresponds to the geometry of the monkey's discriminatory behaviour. If we assume that the monkey has visual experiences, this evidence suggests not only a causal relationship between individual cells and experiences, but also a systematic mapping from spatial structures in the brain to perceived phenomenal structure. In principle, such experiments could be done on human subjects capable of reporting their experiences.

Paul Churchland (1986) has outlined yet another example. The basic idea—the 'state-space sandwich'—arose in neuroscientific work on various 2D-to-2D sensory mappings (coordinate transformations) effected by interconnected sheets of neurones in the brain (see, for example, Pellionisz and Llinas 1985). Churchland speculatively generalizes these ideas to other psychological domains, including the nice discrimination of tastes and colours. For instance, he defines a four-dimensional 'taste-space', whose points represent specific distributions of four neuronal spiking frequencies in a system of four distinct fibres. Crucially, they also represent specific tastes as experienced. He is not describing a mere collection of unrelated correlations between individual neural events and taste experiences. For the abstract state-space represents metrical and similarity relations between the individual points within it, which in turn can be interpreted either as specific neural events or as specific tastes. The hypothesis is that people's introspective reports on the similarities and dissimilarities between different tastes will map onto the similarity-metric implicit in the scientifically defined taste-space. In that case, we could be in a position to predict that a newly discovered (never-tasted) substance, with such-and-such observable effects on the relevant neural mechanisms, will taste very like x, something like y, and not at all like z—indeed, we could say that it must taste like this.

The four dimensions of taste-space are based on the four types of taste receptors found in the tongue. But there is nothing special about the number four. If the neuroscientific evidence suggested that n basic variables were relevant for a certain area of phenomenology, then we could construct an

n-dimensional state-space exhibiting the similarity relations within that domain. Of course, experiments would be needed to confirm that the structure of conscious phenomenology in that domain does indeed map onto the objectively defined state-space. But if it did, we would have an intelligible explanation of why certain experiences seem similar or dissimilar to others, and why a specific brain state causes one particular experience rather than another.

As Churchland admits, the method of coordinate transformations is better suited to computing sensory discriminations and sensorimotor coordinations than to structured linguistic understanding or voluntary action. Different types of neural mechanism, effecting different types of computation, are very likely involved in the so-called 'higher mental processes'.

Current ideas from artificial intelligence (AI), including both connectionism and von Neumann computation, may hold some clues, and are guiding work in computational neuroscience (Churchland and Sejnowski 1992). They also suggest some aspects of what Nagel (1974) calls an 'objective phenomenology', whereby the subjective character of experiences could be described (at least in its structural aspects) in a form comprehensible to beings incapable of having those particular experiences. For example, AI-work on vision (Ullman 1979) has shown that computation of shape is not necessary for computation of identity, which supports the view that a creature could experience object-identity without being able to recognize shape. But these are mere straws in the wind. We shall need many new ideas, at both biological and computational levels, to develop a satisfactory neuroscience.

Churchland's aim, in his theory of state-spaces, is to show that eliminative materialism is conceivable, by describing a method of neural computation that illustrates how it might be realized. He claims that ordinary people will eventually identify their experiences in scientific terms: phenomenal language having become redundant, qualia will be eliminated from our ontology. When the Victorian Alice in Lewis Carroll's books drank from the bottle labelled 'DRINK ME', she described the taste as 'very nice ... a sort of mixed flavour of cherry-tart, custard, pineapple, roast turkey, toffee, and hot buttered toast'. Some future Alice, if Churchland is right, might spontaneously describe it merely by identifying a point in a space defined in terms of specific neural circuits and spiking frequencies.

Many people will baulk at this point, saying that it is inconceivable that we should ever substitute objective scientific terminology for subjective introspective report, because brain and mind are irreducibly different. They may add that we can obviously imagine different people experiencing different tastes, or different colours, even though their brains and sensory organs are physiologically and functionally identical. For that matter, they may say, we can imagine physiologically normal people having no experiences at all. In short, both eliminative materialism and functionalism are inconceivable.

What is conceivable (and what is 'obvious'), however, can depend upon one's scientific knowledge. Biologically ignorant people may think that they can imagine a mermaid. If by this they mean a viable creature with not only the external appearance of half-woman-half-fish, but also the internal anatomy of half-woman-half-fish, then they are mistaken. Knowledge of comparative anatomy (not to mention physiology and biochemistry) forbids us to posit such a creature. The two sets of nerves, blood vessels, and respiratory organs cannot be combined at the mid-plane to give a biologically credible organism. Likewise, it may be our current neuroscientific ignorance that tempts us to believe that the inverted (red-for-green) spectrum is possible, or that Alice (a physiologically normal child) could conceivably experience the taste of turkey whenever the rest of us experience acid drops. If only we knew the biological and computational facts underlying taste and colour vision, we might see that these things are not conceivable, after all.

We might even understand that conscious phenomenology as such (never mind its specific quality) necessarily accompanies certain neural events. In that case, neuroscience would be the solvent causing the problem of other minds to dissolve without remainder.

This last step, I must confess, is one I cannot foresee with any confidence. I believe that the specific qualities of experience may one day be scientifically explained, perhaps along the sorts of lines already sketched above. And certain puzzling facts about self-consciousness, too, are explicable by science (see section 4). The reflexiveness of (some) consciousness is not a major problem: but its subjectivity is. Like Dennett's (1991) fall-guy Otto, I do not see how an objective neuroscience could account for the subjective experience of qualia, or even acknowledge their existence. For that reason, I regard the title of his book—*Consciousness explained*—as inviting prosecution under the Trade Descriptions Act. For that reason, too, I see Penrose's remarks about quantum physics and microtubules as fundamentally problematic.

The existence of consciousness as such remains a mystery—at least, given our present state of knowledge. This is not to deny that some future science might explain it. But this could not be done merely by adding new details, or even new categories, to our current neuroscience. An adequate explanation of the cerebral basis (and the evolutionary origin) of consciousness would involve such a radical shift in our contemporary scientific and philosophical assumptions that we can have no idea, at present, what such an explanation might be like. In this sense, then, I agree with Fodor that we have not the slightest idea how anything material could be conscious.

4. Human identity

The terms 'conscious' and 'consciousness' have many different meanings (Wilkes 1988; Block 1995). One of these concerns self-consciousness. By this I mean the reflexive activity in which one represents oneself as a person, to whom intentional predicates of belief and desire, and their many cognates, are applicable. We need to understand how people apply these mental predicates to themselves, considered as the origin of action (and consciousness) associated with a particular human body—acknowledged as 'their own' body.

Very little is known, at present, about the cerebral basis of human identity. A few hints are found in developmental psychologists' studies of children's ability to use psychological concepts. Beyond that, some clues are provided by abnormal misapplications and/or 'disownings' of intentional and bodily characteristics. Full-blooded intentional concepts seem to be lacking in autism, which, in turn, may be due to a specific brain deficit. Some indications of the criteria used to group beliefs and desires as those of 'one' person are found in abnormal psychology, especially multiple personality disorder. And some cerebral circuits involved in conscious 'ownership' of one's own body are suggested by cases of brain damage in which patients fail to acknowledge as 'theirs' an arm, a leg, or even half of the entire body (Marcel 1993).

Intentional descriptions can be applied to oneself on the basis of behavioural interpretation and conscious introspection. This reflexive activity, being a special case of the intentional stance, favours self-descriptions showing cognitive-teleological coherence. Intentional narratives in general unify (describe, explain, and predict) a person's biography, and self-ascribed narratives unify one's autobiography—and, to some extent, one's life-plans and future behaviour. So Dennett (1991), for instance, characterizes human identity as 'the centre of narrative gravity'.

There is growing evidence that the ability to apply intentional predicates to other people is a human universal, associated with a specific area of the brain (orbito-frontal cortex). In autism, this ability appears to be defective or even entirely absent (although autistic children may apply such predicates to themselves) (Frith 1989; Baron-Cohen 1995). Most experiments on people's 'theory of mind' are purely psychological: they explore subjects' descriptions of and behaviour towards other people, in fictional or real-life situations. But some very recent brain-imaging studies of normal individuals suggest that the blood supply to orbito-frontal cortex increases when they process intentional verbs, as opposed to non-intentional ones (Baron-Cohen *et al.* 1994). Given that our concept of self rests on the reflexive application of intentional concepts, this area of the brain may be differentially involved here also. (There is long-standing clinical evidence that damage to this part of the brain can lead to a failure to introspect, and a lack of self-reflective capacity (Luria 1969; Blumer and Benson 1975).)

Beyond that, we can say very little about the neural mechanisms underlying human identity. Recent work in developmental psychology suggests that the ability to represent one's skills at successively higher levels is required not only for flexible problem-solving but also for conscious awareness, and is possessed only by human beings (Karmiloff-Smith 1992). In computational terms, this 'representational redescription' is described as a movement from procedural to declarative knowledge. Knowledge that was formerly implicit in some behavioural skill becomes explicitly represented at a higher level, and as such is amenable to alteration and (eventually) conscious manipulation. The self-examination involved in complex examples of introspection and moral choice requires one to be able to consider—and to alter—one's own actions. To this extent, then, the theory of representational redescription may be relevant to our theme. But we have only very sketchy ideas, as yet, about computational systems capable of generating explicit representations of their own behaviour (Clark and Karmiloff-Smith 1993); and these computational models do not map onto specific neural mechanisms.

Something can be said, however, albeit in very broad terms, about the computational basis of personal identity. This approach also helps to explain the striking dissociations of consciousness typical of 'multiple personality' (Boden 1994).

Many of us have had the 'self-distancing' experience of reading a youthful diary entry, almost feeling that someone else must have composed this embarrassing nonsense, written though it is in one's own past handwriting. In clinical cases of multiple personality, the self-distancing is much more pronounced, and affects ongoing behaviour and consciousness.

There may be two (or more) memory-streams associated with one and the same physical body, two (or more) internally consistent sets of motivations and beliefs, and two (or more) distinct streams of consciousness. Some conscious experiences may be 'shared', but, if so, the co-consciousness is often non-reciprocal: one motivational stream has access to all the experiences associated with the other, but not vice versa. The repudiation of 'one' personality by the 'other' is not only apparently sincere (sometimes expressed as total ignorance of the 'other's' existence), but often passionate. 'One' personality will refer to 'another' using third-person pronouns, and perhaps a different name, and deny any responsibility for 'her' actions or opinions— even though she may appear to have quasi-telepathic access to 'her' thought and feelings.

We are all sometimes aware of acting 'out of character', doing something inconsistent with our autobiographical narrative and self-image. And we have all changed our minds, tearing up on Tuesday an unwise letter we wrote on Monday. In multiple personality, the anomalous behaviour is more protracted, and its internal teleological coherence greater. Accordingly, the self-description distinguishes various personal narratives, each with its own origin of consciousness. The writing of unwise letters is described using third-person pronouns, instead of the first-person pronouns of normal self-criticism.

We can get some sense of how these extraordinary clinical phenomena are possible, and of how ordinary self-consciousness is constructed also, by thinking of the mind in computational terms. We need to ask what sort of mental architecture could support complex cognitive-motivational systems, and what criteria of teleological coherence are used in constructing a personal narrative.

A human mind includes many motives, often competing for the person's attention and for the use of their hands and time. Questions therefore arise about how limited mental attention and bodily resources can be allocated between the various motives, and how these motives can be prioritized and scheduled. We must distinguish motives of differing urgency, insistence, and importance: speaking of 'strong' and 'weak' motivation is not enough. Compare, for instance, the urgency of a motive to buy bread just as the bakery is about to shut, the insistence of a motive to locate a shop selling one's favourite type of bread, and the importance of a motive to eat food every day.

These distinctions, and many others, are made in Aaron Sloman's account of multiple motivation (Beaudoin and Sloman 1993; Sloman 1990). Sloman discusses a control system for scheduling motives in a teleologically consistent way, one which is sensitive also to mere preferences, to changes in belief, to appropriate emotions, such as grief (Wright *et al.* 1996), and to shifting moods. His theory cannot yet be fully implemented in a computer model. But it helps us to think clearly about what is involved in choosing between our many aims and preferences.

Sloman outlines mechanisms for selecting between competing motives, mechanisms that take seriously distinctions such as those listed above. Detailed problem-solving does not happen whenever there is a motivational clash. On the contrary, an urgent motive is one that has to be satisfied quickly, so that careful consideration of evidence and alternatives would be out of place. If the bakery is about to shut, you have no time to luxuriate in thinking just what sort of bread you want to buy: simply, you must get in there fast. If the motive is both urgent and important (escaping from an approaching tiger), the time available for thought is even shorter.

On the other hand, an important but non-urgent motive may continually be placed at the head of a priority queue, for action or for deliberation. (The more insistent the move is, the more often this will happen.) Accordingly, when the system has no more urgent need, it will consider how the relevant goals might be achieved. This process may require complex problem-solving, involving evaluations of various kind (such as personal preferences and moral codes), inference from stored beliefs, means-end analysis, and contingency-planning. (Sloman's broad-brush picture is neutral as between classical and connectionist AI: the mental architecture must enable these types of function to occur, whatever their detailed implementation may be.)

On such a view, conscious choice is computationally complex. Sometimes, it involves the person's self-image and moral priorities. Even 'unthinking' actions may be part-generated by habitual evaluations that could (if challenged) be consciously affirmed or rejected. Evaluations and beliefs can be altered to varying degrees by reflective thought, and/or by conversation with others. Reflex responses and habitual behaviour can be altered with great difficulty, if at all.

These real differences in the nature of self-control underlie the everyday notion of human freedom. Only agents with significant mental complexity are capable of free choice. Although random choice-points may sometimes occur, the notion that randomness (as opposed to structure) is the core of freedom is mistaken. 'Explanations' of consciousness (and freedom) that insist on randomness should therefore be treated with scepticism, unless the randomness is carefully situated with respect to some relevant computational structure. (Penrose stresses that his account of quantum activity implies that 'whatever happens is effectively random' (1994b, p. 248), but says nothing about the structural differences between decisions of various sorts.)

In normal people (and in many computer systems), two or more motives may be approached, or even achieved, by the same activity. A system that wanted many different things, but could pursue only one goal at a time, would be unable to do this. Control of its behaviour would flit from one motive to another, appearing excessively single-minded over brief periods of time. There would be much wasted effort: not only unnecessary repetitions (due to its not being able to kill two birds with one stone), but self-defeating actions too, wherein a goal or sub-goal that has already been achieved is later undone in pursuit of some quite different end. The similarity of such behaviour to certain aspects of multiple personality is evident.

Mental dissociation—in consciousness or behaviour—results from various types of compromise or breakdown in the complex control system that informs and unifies the normal mind (whose motivational consistency is by no means perfect). Independent, alternating, and perhaps even competing motivational structures will very likely arise if the usual control mechanisms for integrating motives break down. These teleologically coherent structures are candidates for distinct personal narratives. Differential access to memory, which plays an important criterial role in individuating 'personalities' or 'alternates', could also arise in this way. If we think of the mind as a computational system, we can see how it is possible for some memories to be shared between several (or all) alternates, and for others to be accessible only to one (or two ...). The occurrence of reciprocal and non-reciprocal co-consciousness, too, can be understood in these terms.

It seems implausible that damage to a highly specific site in the brain is responsible for multiple personality. Rather, the condition seems to be sometimes an artefact of suggestion, and sometimes a spontaneous functional disorder prompted by childhood abuse (the dissociation serving to reduce both personal vulnerability to and responsibility for the abuse). But some other cases of breakdown in the ownership of action do appear to be caused, at least in part, by identifiable lesions in the brain. Hemi-agnosic patients, for instance, seem unaware of one half of their body—or, rather, do not recognize it as their own (Bisiach 1993; Bisiach and Berti 1994). The arm lying on the bed is not 'their' arm, one side of the face in the mirror does not merit make-up, and actions done by the neglected side of the body are disowned. Again, both brain-imaging and clinical observation suggest that auditory hallucinations in schizophrenia often occur when the person's own speech is not recognized as their own, so is attributed by them to someone else (Frith 1992).

Although we lack a detailed scientific explanation of human identity, we do have some idea of how the integration of normal self-consciousness might be achieved, and how it might give way to the conscious dissociations of multiple personality. We even have some preliminary evidence that a particular part of the brain is involved in the ascription of intentional predicates. With respect to human identity, then, we are in much the same position as we were when considering consciousness of other sorts. Scientific enquiry can help us to understand why consciousness of *this* type rather than *that* occurs on a given occasion. But why any consciousness should occur at all remains a mystery.

5. Questioning Cartesianism

I noted in section 2 that recent neuroscience makes many claims with which Descartes would have been in sympathy. And in sections 3 and 4, I asked how those claims might be made less starkly counterintuitive. Some philosophers, however, regard mind–brain unintelligibility as an inevitable result of a fundamental philosophical mistake, inherited by neuroscience from Descartes.

Neuroscientists typically adopt the Cartesian assumption that there is a fundamental distinction between subjective consciousness and objective reality. A corollary of that assumption is that the aim of science is to discover the nature of that reality. Since the human mind is limited in various ways, there may be aspects of external reality that are inaccessible to our senses and beyond the reach of our measuring instruments. If so, we cannot discover them. But they exist nonetheless, independent of human minds. In short, science is a realist enterprise.

Some philosophers take a very different position, according to which there is no objective reality independent of subjective knowledge of it. It follows that science is not the study of independently existing objects. To do science is to take up an objective attitude, to attempt to construct knowledge in a particular way—one that allows of intersubjective agreement. And science can never explain the origin of subjectivity, since the scientific attitude is itself grounded in it.

Versions of this view are found in Wittgensteinian philosophy and continental phenomenology. These philosophical ideas have been used in criticisms of current scientific projects aimed at understanding the mind. Hubert Dreyfus (1979), for instance, draws heavily on Heidegger and Wittgenstein in arguing that even bodily skills (never mind consciousness) cannot be scientifically explained. Although his arguments are primarily directed against AI, they cast doubt also on the possibility of a neurophysiological explanation of human behaviour and experience. Remarks of the form 'We just do it' are legion in Dreyfus' writing.

Anti-Cartesianism is not confined to people unsympathetic to science. Attacks on the subject–object distinction, and the representational psychology that accompanies it, have recently surfaced in cognitive science—notably, in the study of adaptive behaviour and situated robotics (Wheeler, 1996). On this view, 'subject' and 'environment' should be mutually defined. It is not enough to say (what a Cartesian cognitive science would admit) that different animals inhabit different 'worlds' in the sense that their sensory-motor capacities enable them to respond only to selected aspects of the objective world. Nor is it enough to point out (as ecological psychologists do) that the environment offers different 'affordances' to different species, according to their sensory-motor equipment and behavioural repertoire. Rather, adaptation to and construction of the environment go hand in hand, and cannot be clearly distinguished. Ethology and psychology should therefore focus on how situated, subject-engendered, meanings emerge from the bodily actions of creatures embedded in their self-generated world.

The neuroscientific approach associated with this position describes the brain and its parts as dynamical systems (Thelen and Smith 1993; Port and van Gelder 1995). Indeed, organism and environment are seen as closely coupled dynamical systems. We may find it convenient to speak of 'two' mutually interacting systems (at a given level of description), but these can always be thought of as one larger system. Moreover, neural mechanisms whose activity is correlated with specific environmental features are described in a non-Cartesian way, with no reference to representation or computation. Orientation-detectors, on this view, do not represent surface contours, nor do they compute information about orientation. Rather, these visual neurones are dynamically coupled with physical contours (and with motor systems in the body) in a way that enables the organism to discriminate edges oriented in different ways.

'Discriminate', here, implies discriminative behaviour. Whether it implies conscious experience also is problematic. The experiments described in section 3 were done on the (Cartesian) assumption that there are two fundamentally different types of event, neural processes and conscious experience, which may turn out to be systematically correlated. Only after such correlations are discovered do Cartesian philosophers suggest that the 'two' events are identical. But there can be no correlations between brain processes and subjectivity understood in the non-Cartesian sense. That subjectivity is the ground of all neuroscientific categories, so can hardly be 'correlated' with any of them. Indeed, it is the ground of all categories whatsoever. The status of the 'consciousness' studied in psychophysiological experiments is thus subtly shifted. It is no longer epistemologically fundamental but, like 'objects', derivative. Even if a scientific explanation of it were constructed, the more fundamental form of subjectivity—in which all our knowledge is grounded—would remain untouched.

6. Conclusion

If one asks whether we shall ever have a complete science of consciousness, the first answer must be 'Perhaps not'. But this is relatively uninteresting, because we must give the same answer to the question whether we shall ever have a complete understanding of kidney function. To have a complete understanding, all relevant questions would need to have been correctly answered. We can never be confident that all potentially relevant questions have even been asked, never mind answered. This is not merely a matter of failing to discover the most recondite and elusive details, but also of allowing for future changes in explanatory strategy. The fate of Kant's Newtonian metaphysics shows how philosophers may be misled if they assume that current theories are the last word in science.

A more interesting, if more depressing, answer might be 'No: it is potentially there, but we are not capable of finding it'. Colin McGinn (1991) believes that where consciousness (rather than kidneys) is concerned, aspiring scientists are in a hopeless situation. According to him, the mind–body problem is insoluble—or, rather, it is insoluble by human beings. He does not doubt that consciousness is wholly dependent on the brain, with no magic or mysticism involved. Nor does he seek to deny that science gives us knowledge of reality. But the link between brain and consciousness, he says, can be glimpsed by us only in a superficial way (in terms of correlations, for example). Humans are constitutionally incapable of generating the scientific concepts that would be required to understand the mind–brain relation in a more fundamental way, much as dogs are inherently incapable of understanding evolutionary theory or quantum physics.

This may, conceivably, be true. But at present, *pace* McGinn, we have no convincing reason to think so. Our scientific knowledge of brain mechanisms is so scanty, and the history of neuroscience so short, that it would be defeatist to give up now.

My reply to McGinn assumes, as he does himself, that consciousness is somehow generated by a physical system. But perhaps that assumption is mistaken—indeed, absurd. The fundamental question here seems to be whether we should accept or reject the Cartesian subject–object distinction, with its associated view that science in general is knowledge of objective realities.

If we accept it, then a future neuroscience may be able to explain why this conscious experience rather than that one occurred on a given occasion. But the occurrence of subjective consciousness as such may well remain inexplicable. If we reject it, we reject scientific realism also, and put human subjectivity in a position where it is prior to science, not something to be studied by it. In that event, the philosophical status of the 'mind–body correlations' already discovered seems problematic.

Perhaps Fodor is right, after all.

References

Baron-Cohen, S. (1995). *Mindblindness: an essay on autism and theory of mind.* MIT Press, Cambridge, Mass.

Baron-Cohen, S., Ring, H., Moriarty, J., Schmitz, B., Costa, D., and Ell, P. (1994). Recognition of mental state terms: clinical findings in children with autism and a functional neuroimaging study of normal adults. *British Journal of Psychiatry*, **165**, 640–9.

Beaudoin, L. P. and Sloman, A. (1993). A study of motive processing and attention. In *Prospects for artificial intelligence* (ed. A. Sloman, D. Hogg, G. Humphreys, D. Partridge and A. Ramsay), pp. 229–38. IOS Press, Amsterdam.

Bisiach, E. (1993). Mental representation in unilateral neglect and related disorders. *Quarterly Journal of Experimental Psychology*, **47A**, 435–61.

Bisiach, E. and Berti, A. (1994). Consciousness in dyschiria. In *The cognitive neurosciences* (ed. M. Gazzaniga), pp. 1331–46. MIT Press, Cambridge, Mass.

Block, N. (1995). On a confusion about a function of consciousness (with peer-review). *Behavioral and Brain Sciences*, **18**, 227–87.

Blumer, D. and Benson, D. (1975). Personality changes with frontal and temporal lobe lesions. In *Psychiatric aspects of neurological disease* (ed. D. Blumer and D. Benson), Vol. 1, pp. 151–70. Grune & Stratton, New York.

Boden, M. A. (1994). Multiple personality and computational models. In *Philosophy, psychology, and psychiatry* (ed. A. Phillips-Griffiths), pp. 103–14. Cambridge University Press, Cambridge, 1994.

Churchland, P. M. (1979). *Scientific realism and the plasticity of mind.* Cambridge University Press, Cambridge.

Churchland, P. M. (1990). *A neurocomputational perspective: the nature of mind and the structure of science.* MIT Press, Cambridge, Mass.

Churchland, P. M. (1986). Some reductive strategies in cognitive neurobiology. *Mind*, **45**, 279–309. (Reprinted in P. M. Churchland, *A neurocomputational perspective.*)

Churchland, P. S. and Sejnowski, T. (1992). *Computational neuroscience.* MIT Press, Cambridge, Mass.

Clark, A. and Karmiloff-Smith, A. (1993). The cognizer's innards: a psychological and philosophical perspective on the development of thought (with peer-commentary and reply). *Mind and Language*, **8**, 487–519.

Crick, F. H. C. (1994). *The astonishing hypothesis: the scientific search for the soul.* Macmillan, New York.

Crick, F. H. C. and Koch, C. (1990). Towards a neurobiological theory of consciousness. *Seminars in Neuroscience*, **2**, 263–75.

Dennett, D. C. (1988). Quining qualia. In *Consciousness in contemporary science* (ed. A. Marcel and E. Bisiach), pp. 42–77. Oxford University Press, Oxford.

Dennett, D. C. (1991). *Consciousness explained.* Little Brown, Boston, Mass.

Dreyfus, H. (1979). *What computers can't do*, (2nd edn). Harper & Row, New York.

Fodor, J. A. (1992). The big idea: can there be a science of mind? *The Times Literary Supplement.*

Frith, C. (1992). *The cognitive neuropsychology of schizophrenia.* Erlbaum Press, Hove.

Frith, U. (1989). *Autism: explaining the enigma.* Blackwell, Oxford.

Grush, R. and Churchland, P. S. (1995). Gaps in Penrose's toilings. *Journal of Consciousness Studies*, **2**, 10–29.

Hubel, D. (1988). *Eye, brain, and vision.* W. H. Freeman, Oxford.

Huxley, T. H. (1893). On the hypothesis that animals are automata, and its history. In *Method and results: essays* (by T. H. Huxley), pp. 199–250. Macmillan, London.

Jackson, F. (1982). Epiphenomenal qualia. *Philosophical Quarterly*, **32**, 127–36.

Karmiloff-Smith, A. (1992). *Beyond modularity: a developmental perspective on cognitive science.* MIT Press, Cambridge, Mass.

Logothetis, N. and Schall, J. (1989). Neuronal correlates of subjective visual perception. *Science*, 245, 761–3.

Luria, A. (1969). Frontal lobe syndromes. In *Handbook of clinical neurology* (ed. P. Vinken and G. Bruyn), Vol. 2, pp. 725–57. North Holland, Amsterdam.

McGinn, C. (1991). *The problem of consciousness.* Blackwell, Oxford.

Marcel, A. J. (1993). Slippage in the unity of consciousness. In *Experimental and theoretical studies of consciousness*, CIBA Foundation Symposium 174 (ed. G. R. Bock and J. Marsh), pp. 168–86. Wiley, Chichester.

Milner, B. and Rugg, M. (ed.) (1992). *The neuropsychology of consciousness*. Academic Press, New York.

Nagel, T. (1974). What is it like to be a bat? *Philosophical Review*, **83**, 435–50.

Newsome, W. T. and Salzman, C. D. (1994). The neuronal basis of motion perception. In *Experimental and theoretical studies of consciousness*, CIBA Foundation Symposium 174 (ed. G. R. Bock and J. Marsh), pp. 217–30. Wiley, Chichester.

Pellionisz, A. and Llinas, R. (1985). Tensor network theory of the metaorganization of functional geometries in the central nervous system. *Neuroscience*, **16**, 245–74.

Penrose, R. (1989). *The emperor's new mind*. Oxford University Press, Oxford.

Penrose, R. (1994a). *Shadows of the mind: a search for the missing science of consciousness*. Oxford University Press, Oxford.

Penrose, R. (1994b). Mechanisms, microtubules, and mind. *Journal of Consciousness Studies: Controversies in Science and the Humanities*, **1**, 241–9.

Penrose, R. and Hameroff, S. (1995). *Journal of Consciousness Studies*, **2**, 99–112.

Place, U. T. (1956). Is consciousness a brain process? *British Journal of Psychology*, **67**, 44–50.

Port, R. and van Gelder, T. (ed.) (1995). *Mind as motion: dynamics, behavior, and cognition*. MIT Press, Cambridge, Mass.

Posner, M. I. and Raichle, M. E. (1994). *Images of mind*. Walter Freeman, San Francisco.

Putnam, H. (1960). Minds and machines. In *Dimensions of mind* (ed. S. Hook), pp. 56–69. Collier, New York.

Putnam, H. (1967). The nature of mental states. Reprinted in H. Putnam (1975). *Mind, language, and reality: philosophical papers*, Vol. 2, pp. 429–40. Cambridge University Press, Cambridge.

Searle, J. R. (1992). *The rediscovery of the mind*. MIT Press, Cambridge.

Sloman, A. (1990). Motives, mechanisms, and emotions. In *The philosophy of artificial intelligence* (ed. M. A. Boden), pp. 231–47. Oxford University Press, Oxford.

Smart, J. J. C. (1959). Sensations and brain processes. *Philosophical Review*, **68**, 141–56.

Thelen, E. and Smith, L. B. (1993). *A dynamic systems approach to the development of cognition and action*. MIT Press, Cambridge, Mass.

Ullman, S. (1979). *The interpretation of visual motion*. MIT Press, Cambridge.

Watson, J. B. (1913). Psychology as the behaviorist views it. *Psychological Review*, **20**, 158–77.

Weiskrantz, L. (1986). *Blindsight: a case study and implications*. Clarendon Press, Oxford.

Weiskrantz, L. (1990). Outlooks for blindsight: explicit methodologies for implicit processes. *Proceedings of the Royal Society, London*, **B239**, 247–78.

Wheeler, M. W. (1996). From robots to Rothko: the bringing forth of worlds. In *The philosophy of artificial life* (ed. M. A. Boden), pp. 209–36. Oxford University Press, Oxford.

Wilkes, K. V. (1988). - - - - Yishi, Duh, Um, and Consciousness. In *Consciousness in contemporary science* (ed. A. Marcel and E. Bisiach). Oxford University Press, Oxford.

Wright, I. P., Sloman, A., and Beaudoin, L. P. (1996). The architectural basis for grief. *Philosophy, Psychiatry, and Psychology*, **3(2)**, 101–37.

CHAPTER 2

How to study consciousness scientifically

JOHN R. SEARLE

The neurosciences have now advanced to the point that we can address—and perhaps, in the long run, even solve—the problem of consciousness as a scientific problem like any other. There are, however, some philosophical obstacles to this project. The aim of this article is to address and try to overcome some of those obstacles. Because the problem of giving an adequate account of consciousness is a modern descendant of the traditional 'mind–body problem', I will begin with a brief discussion of the mind–body issue.

The mind–body problem can be divided into two problems: the first is easy to solve, the second is much more difficult. The first problem is to explain the general character of the relations between consciousness and other mental phenomena on the one hand and the brain on the other. The solution to this easy problem can be given with two principles: (1) consciousness and indeed all mental phenomena are caused by lower-level neurobiological processes in the brain; (2) consciousness and other mental phenomena are higher-level features of the brain. I have expounded this solution to the mind–body problem in several writings, so I won't say more about it here.[1]

The second and more difficult problem is to explain in detail how consciousness actually works in the brain. I believe that a solution to this problem would be the most-important scientific discovery of the present era. When—and if—it is made it will answer the following question: 'How exactly do neurobiological processes in the brain cause consciousness?' Given our present models of brain functioning, it would answer the question, 'How exactly do the lower-level neuronal firings at synapses cause all of the enormous variety of our conscious (subjective, sentient, aware) experiences?' Perhaps we are wrong to think that neurones and synapses are the right anatomical units to account for consciousness, but we do know that some elements of brain anatomy must be the right level of description for answering our question. We know that because we know that brains cause consciousness, in a way that elbows, livers, television sets, cars, and commercial computers do not. Therefore the special features of brains—features that they do not have in common with elbows, livers, etc.—must be essential to the causal explanation of consciousness.

The explanation of consciousness is essential for explaining most of the features of our mental life because in one way or another they involve consciousness. How exactly do we have visual and other sorts of perceptions? What exactly is the neurobiological basis of memory, and of learning? What are the mechanisms by which nervous systems produce sensations of pain? What, neurobiologically speaking, are dreams and why do we have them? Even: why does alcohol make us drunk and why does bad news make us feel depressed? In fact I do not believe we can have an adequate understanding of *unconscious* mental states until we know more about the neurobiology of consciousness.

As I said at the beginning, our ability to get an explanation of consciousness—a precise neurobiology of consciousness—is in part impeded by a series of philosophical confusions. This is one of those areas of science (and they are more common than you might suppose) where scientific progress is blocked by philosophical error. And because many scientists and philosophers make these errors, I shall devote this article to trying to remove what I believe are some of the most serious philosophical obstacles to understanding the relation of consciousness to the brain.

It may seem presumptuous for a philosopher to try to advise scientists in an area outside his special competence, so I want to begin by making a few remarks about the relation of philosophy to science and about the nature of the problem we are discussing. 'Philosophy' and 'science' do not name distinct subject matters in the way that 'molecular biology', 'geology', and 'the history of Renaissance painting' name distinct subject areas; rather, at the abstract level at which I am now considering these issues, there is no distinction of subject matter because, in principle at least, both philosophy and science are universal in subject matter. And of the various parts of this universal subject matter, each aims for knowledge. When knowledge becomes systematic we are more inclined to call it scientific knowledge, but knowledge as such contains no restriction on subject matter. 'Philosophy' is in large part the name for all those questions that we do not know how to answer in the systematic way that is characteristic of science. These questions include, but are not confined to, the large family of conceptual questions that have traditionally occupied philosophers: the nature of truth, justice, knowledge, meaning, etc.

For the purposes of this discussion, therefore, the only important distinction between philosophy and science is this: science is systematic knowledge; philosophy is in part an attempt to get us to the point where we can have systematic knowledge. That is why science is always right and philosophy is always wrong: as soon as we think we really know something we stop calling it philosophy and start calling it science. Beginning in the seventeenth century the area of scientific knowledge increased with the growth of systematic methods for acquiring knowledge. Most of the questions that most bother us, however, have not yet been amenable to the methods of scientific investigation. But we do not know how far we can go with those methods and we should be reluctant to say a priori that such and such questions are beyond the reach of science. I will have more to say about this issue later, because many scientists and philosophers think that the whole subject of consciousness is somehow beyond the reach of science.

A consequence of these points is that there are no 'experts' in philosophy in the way that there are in the sciences. There are experts on the history of philosophy and experts in certain specialized corners of philosophy, such as mathematical logic, but on most of the central philosophical questions there is no such thing as an established core of expert opinion. I remark on this because I often encounter scientists who want to know what philosophers think about a particular issue. They ask these questions in a way that suggest that they think there is a body of expert opinion that they hope to consult. But in the way that there is an answer to the question, 'What do neurobiologists currently think about LTP (long-term potentiation)?', there is no comparable answer to the question, 'What do philosophers currently think about consciousness?' Another consequence of these points is that you have to judge for yourself whether what I have to say in this article is true. I cannot appeal to a body of expert opinion to back me up. If I am right, what I say should seem obviously true, once I have said it and once you have thought about it.

The method I will use in my attempt to clear the ground of various philosophical obstacles to the examination of the question, 'How exactly do brain processes cause consciousness?', is to present a series of views that I think are false or confused and then, one by one, try to correct them by explaining why I think they are false or confused. In each case I will discuss views I have found to be widespread among practising scientists and philosophers.

Thesis 1

Consciousness is not a suitable subject for scientific investigation because the very notion is ill-defined. We do not have anything like a scientifically acceptable definition of consciousness and it is not easy to see how we could get one, because consciousness is unobservable. The whole notion of consciousness is at best confused and at worst it is mystical.

Answer to thesis 1

We need to distinguish analytic definitions, which attempt to tell us the essence of a concept, from common-sense definitions, which just make clear what we are talking about.

An example of an analytic definition is:

Water = df. H_2O

A common-sense definition of the same word is, for example:

- Water is a clear, colourless, tasteless liquid. It falls from the sky in the form of rain, and it is the liquid in lakes, rivers, and seas.

Notice that analytic definitions typically come at the end not at the beginning of a scientific investigation. What we need at this point in our work is a common-sense definition of consciousness, such a definition is not hard to give:

- 'Consciousness' refers to those states of sentience or awareness that typically begin when we wake from a dreamless sleep and continue through the day until we fall asleep again, die, go into a coma, or otherwise become 'unconscious'. Dreams are also a form of consciousness, though in many respects they are quite unlike normal waking states.

Such a definition, whose job is to identify the target of scientific investigation and not to provide an analysis, is adequate and indeed is exactly what we need to begin our study. Because it is important to be clear about the target, I want to note several consequences of the definition:

1. Consciousness, so defined, is an inner qualitative, subjective state typically present in humans and the higher mammals. We do not at present know how far down the phylogenetic scale it goes, and until we get an adequate scientific account of consciousness it is not useful to worry about whether, for example, snails are conscious.

2. Consciousness, so defined, should not be confused with attention, because there are many things I am conscious of that I am not paying attention to—such as the feeling of the shirt on my back.

3. Consciousness, so-defined, should not be confused with self-conscious-ness. Consciousness, as I am using the word, refers to any state of sentience or awareness, but self-consciousness, in which the subject is aware of himself or herself, is a very special form of consciousness, perhaps peculiar to humans and the higher animals. Forms of consciousness such as feeling a pain do not necessarily involve a consciousness of a self as a self.

4. I experience my own conscious states, but I can neither experience nor observe those of another human or animal, nor can they experience or observe mine. But the fact that the consciousness of others is 'unobservable' does not by itself prevent us from getting a scientific account of consciousness. Electrons, black holes, and the Big Bang are not observable by anybody, but that does not prevent their scientific investigation.

Thesis 2

Science is, by definition, objective, but on the definition of consciousness you have provided you admit it is subjective. So, it follows from your definition that there cannot be a science of consciousness.

Answer to thesis 2

I believe that this statement reflects several centuries of confusion about the distinction between objectivity and subjectivity. It would be a fascinating exercise in intellectual history to trace the vicissitudes of the objective/subjective distinction. In Descartes's writings in the seventeenth century, 'objective' had something close to the opposite of its current meaning.[2] Sometime—I don't know when—between the seventeenth century and the present, the objective–subjective distinction rolled over in bed.

For present purposes, however, we need to distinguish between the epistemic sense of the objective–subjective distinction and the ontological sense. In the epistemic sense, objective claims are objectively verifiable or objectively knowable, in the sense that they can be known to be true or false in a way that does not depend on the preferences, attitudes, or prejudices of particular human subjects. So, if I say, for example, 'Rembrandt was born in 1606', the truth or falsity of that statement does not depend on the particular attitudes, feelings, or preferences of human subjects. It is, as they say, a matter of objectively ascertainable fact. This statement is epistemically objective. It is an objective fact that Rembrandt was born in 1606.

This statement differs from subjective claims whose truth cannot be known in this way. So, for example, if I say 'Rembrandt was a better painter than Rubens', that claim is epistemically subjective, because, as we would say, it is a matter of subjective opinion. There is no objective test, nothing independent of the opinions, attitudes, and feelings of particular human subjects, that would be sufficient to establish that Rembrandt is a better painter than Rubens.

I hope the distinction between objectivity and subjectivity in the epistemic sense is intuitively clear. But there is another distinction that is related to the epistemic objective–subjective distinction but should not be confused with it;

that is, the distinction between ontological objectivity and subjectivity. Some entities have a subjective mode of existence; some have an objective mode of existence. So, for example, my present feeling of pain in my lower back is ontologically subjective in the sense that it only exists as experienced by me. In this sense, all conscious states are ontologically subjective, because they have to be experienced by a human or an animal subject to exist. In this respect, conscious states differ from, for example, mountains, waterfalls, or hydrogen atoms. Such entities have an objective mode of existence, because they do not have to be experienced by a human or animal subject to exist.

Given this distinction between the *ontological* sense of the objective–subjective distinction, and the *epistemic* sense of the distinction, we can see the ambiguity of the claim made in thesis 2. Science is indeed objective in the epistemic sense. We seek truths that are independent of the feelings and attitudes of particular investigators. It does not matter how you feel about hydrogen, whether you like it or don't like it; hydrogen atoms have one electron. It is not a matter of opinion. That is why the claim that Rembrandt is a better painter than Rubens is not a scientific claim. But, now, the fact that science seeks objectivity in the epistemic sense should not blind us to the fact that there are ontologically subjective entities that are as much a matter of scientific investigation as any other biological phenomena. We can have epistemically objective knowledge of domains that are onto-logically subjective. So, for example, in the epistemic sense, it is an objective matter of fact—not a matter of anybody's opinion—that I have pains in my lower back. But the existence of the pains themselves is ontologically subjective.

The answer, then, to thesis 2 is that the requirement that science be objective does not prevent us from getting an epistemically objective science of a domain that is ontologically subjective.

Thesis 3
There is no way that we could ever give an intelligible causal account of how anything subjective and qualitative could be caused by anything objective and quantitative, such as neurobiological phenomena. There is no way to make an intelligible connection between objective third-person phenomena, such as neurone firings, and qualitative, subjective states of sentience and awareness.

Answer to thesis 3
Of all the theses we are considering, this seems to me the most challenging. In the hands of some authors (e.g., Thomas Nagel[3]) it is presented as a serious obstacle to getting a scientific account of consciousness using anything like

our existing scientific apparatus. The problem, according to Nagel, is that we have no idea how objective phenomena, such as neurone firings, could necessitate, could make it unavoidable, that there be subjective states of awareness. Our standard scientific explanations have a kind of necessity, and this seems to be absent from any imaginable account of subjectivity in terms of neurone firings. For example, what fact about neurone firings in the thalamus could make it necessary that anybody who has those firings in that area of the brain must feel a pain?

Although I think this is a serious problem for philosophical analysis, for the purpose of the present discussion, there is a rather swift answer to it: we know in fact that it happens. That is, we know as a matter of fact that brain processes cause consciousness. The fact that we don't have a theory that explains how it is possible that brain processes could cause consciousness is a challenge for philosophers and scientists. But it is not a challenge to the fact that brain processes do in fact cause consciousness, because we know independently of any philosophical or scientific argument that they do. The mere fact that it happens is enough to tell us that we should be investigating the form of its happening and not challenging the possibility of its happening.

So I accept the unstated assumption behind thesis 3: given our present scientific paradigms it is not clear how consciousness is caused by brain processes. But I see that as analogous to: within the explanatory apparatus of Newtonian mechanics, it is not clear how there could exist a phenomenon such as electromagnetism; within the explanatory apparatus of nineteenth-century chemistry, it is not clear how there could be a non-vitalistic, chemical explanation of life. That is, I see the problem as analogous to earlier apparently unsolvable problems in the history of science. The challenge is to forget about how we think the world ought to work, and instead figure out how it works in fact.

My own guess—and at this stage in the history of knowledge it is only a speculation—is that when we have a general theory of how brain processes cause consciousness, our sense that it is somehow arbitrary or mysterious will disappear. In the case of the heart, for example, it is clear how the heart causes the pumping of blood. Our understanding of the heart is such that we see the necessity. Given these contractions blood must flow through the arteries. What we so far lack for the brain is an analogous account of how the brain causes consciousness. But if we had such an account—a general causal account—then our sense of mystery and arbitrariness would disappear.

It is worth pointing out that our sense of mystery has already changed since the seventeenth century. To Descartes and the Cartesians, it seemed mysterious that a physical impact on our bodies should cause a sensation in our souls. But we have no trouble in sensing the necessity of pain, given certain sorts of impacts on our bodies. We do not think it at all mysterious that the man whose foot is caught in the punch press is suffering terrible pain. We have moved the sense of mystery inside. It now seems mysterious to us that neurone firings in the thalamus should cause sensations of pain. But I suggest that a thorough-going neurobiological account of how and why exactly it happens would remove this sense of mystery.

Thesis 4

All the same, within the problem of consciousness we need to separate out the qualitative, subjective features of consciousness from the measurable objective aspect that can be properly studied scientifically. These subjective features, sometimes called 'qualia', can be safely left on one side. That is, the problem of qualia needs to be separated from the problem of consciousness. Consciousness can be defined in objective third-person terms and the qualia can then be ignored. And, in fact, this is what the best neurobiologists are doing. They separate the general problem of consciousness from the special problem of qualia.

Answer to thesis 4

I would have not have thought that this thesis—that consciousness could be treated separately from qualia—was commonly held until I discovered it in several recent books on consciousness.[4] The basic idea is that problem of qualia can be carved off from consciousness and treated separately or, better still, simply brushed aside. That seems to me profoundly mistaken. There are not two problems, the problem of consciousness and a subsidiary problem, the problem of qualia. *The problem of consciousness is identical with the problem of qualia, because conscious states are qualitative states right down to the ground.* Take away the qualia and there is nothing there. This is why that I never use the word 'qualia', except in sneer quotes, because it suggests that there is something else to consciousness besides qualia, and there isn't. Conscious states by definition are inner, qualitative, subjective states of awareness or sentience.

Of course, it is open to anybody to define these terms as he or she likes and use the word 'consciousness' for something else. But then we would still have the problem of what I am calling 'consciousness', which is the problem of accounting for the existence of our ontologically subjective states of awareness. The point for the present discussion is that the problem of consciousness and the problem of so-called qualia is the same problem; and you cannot evade the identity by treating consciousness as some third-person, ontologically objective phenomena and setting qualia on one side, because to do so is simply to change the subject.

Thesis 5

Even if consciousness did exist, as you say it does, in the form of subjective states of awareness or sentience, all the same it couldn't make a real difference to the real physical world. It would just be some surface phenomenon that didn't matter causally to the behaviour of the organism in the world. In the current philosophical jargon, consciousness would be epiphenomenal. It would be like surface reflections on the water of the lake or the froth on the wave coming to the beach. Science can offer an explanation why there are surface reflections and why the waves have a froth, but in our basic account of how the world works, these surface reflections and bit of froth are themselves caused, but are causally insignificant in producing further effects. Think of it this way. If we were doing computer models of cognition, we might have one computer that performed cognitive tasks, and another one, just like the first, except that the second computer was lit up with a purple glow. Now that is what consciousness amounts to: a scientifically irrelevant, luminous purple glow. And the proof of this point is that for any apparent explanation in terms of consciousness a more fundamental explanation can be given in terms of neurobiology. For every explanation of the form, for example, my conscious decision to raise my arm caused my arm to go up, there is a more fundamental explanation in terms of motor neurones, acetylcholine, etc.

Answer to thesis 5

It might turn out that in our final scientific account of the biology of conscious organisms, the consciousness of these organisms plays only small or negligible role in their life and survival. This is logically possible in the sense, for example, that it might turn out that DNA is irrelevant to the inheritance of biological traits. It might turn out that way but it is most unlikely, given what we already know. Nothing in thesis 5 is a valid argument in favour of the causal irrelevance of consciousness.

There are indeed different levels of causal explanation in any complex system. When I consciously raise my arm, there is a macro level of explanation in terms of conscious decisions, and a micro level of explanation in terms of synapses and neurotransmitters. But, as a perfectly general point about complex systems, the fact that the macro-level features are themselves caused by the behaviour of the micro elements and realized in the system composed of the micro elements does not show that the macro-level features are epiphenomenal. Consider, for example, the solidity of the pistons in my car engine. The solidity of the piston is entirely explainable in terms of the behaviour of the molecules of the metal alloy of which the piston is composed; and for any macro-level explanation of the workings of my car engine given in terms of pistons, the crank shaft, sparkplugs, etc., there will be micro levels of explanation given in terms of molecules of metal alloys, the oxidization of hydrocarbon molecules, etc. But this does not show that the solidity of the piston is epiphenomenal. On the contrary, such an explanation explains why you can make effective pistons out of steel and not out of butter or papier mâché. Far from showing the macro level to be epiphenomenal, the micro level of explanation explains, among other things, why the macro levels are causally efficacious. That is, in such cases the bottom-up, causal explanations of macro-level phenomena show why the macrophenomena are not epiphenomenal. An adequate science of consciousness should analogously show how my conscious decision to raise my arm causes my arm to go up by showing how the consciousness, as a biological feature of the brain, is grounded in the micro-level neurobiological features.

The point that I am making here is quite familiar: It is basic to our world view that higher-level or macro features of th world are grounded in or implemented in micro structures. The grounding of the macro in the micro does not by itself show that the macro phenomena are epiphenomenal. Why then do we find it difficult to accept this point where consciousness and the brain are concerned? I believe the difficulty is that we are still in the grip of a residual dualism. The claim that mental states must be epiphenomenal is supported by the assumption that because consciousness is non-physical, it could not have physical effects. The whole thrust of my argument has been to reject this dualism. Consciousness is an ordinary biological, and therefore physical, feature of the organism, as much as digestion or photosynthesis. The fact that it is a physical biological feature does not prevent it from being an ontologically subjective mental feature. The fact that it is both a higher level and a mental feature is no argument at all that it is epiphenomenal, any more than any other higher-level biological feature is epiphenomenal. To repeat, it might turn out to be epiphenomenal, but no valid a priori philosophical argument has been advanced to show that it must turn out that way.

Thesis 6

Your last claims fail to answer the crucial question about the causal role of consciousness. That question is, what is the evolutionary function of consciousness? No satisfactory answer has ever been proposed to that question, and it is not easy to see how one will be forthcoming because it is easy to imagine beings behaving just like us who lack these 'inner, qualitative, states' you have been describing.

Answer to thesis 6

I find this point very commonly made, but if you think about it I hope you will agree that it is a very strange claim to make. Suppose someone asked, what is the evolutionary function of wings on birds? The obvious answer is that for most species of birds the wings enable them to fly and flying increases their genetic fitness. The matter is a little more complicated because not all winged birds are able to fly (consider penguins, for example) and, more interestingly, according to some accounts, the earliest wings were really stubs sticking out of the body that functioned to help the organism keep warm. But there is no question that relative to their environments most birds are immensely aided by having wings with which they can fly.

Now suppose somebody objected by saying that we could imagine the birds flying just as well without wings. What are we supposed to imagine? That the birds are born with rocket engines? The evolutionary question only makes sense given certain background assumptions about how nature works. Given the way that nature works, the function of bird wings is to enable them to fly. And the fact that we can imagine a science-fiction world in which birds fly just as well without wings is really irrelevant to the evolutionary question. Now similarly with consciousness. The way that human and animal intelligence works is through consciousness. We can easily imagine a science-fiction world in which unconscious zombies behave exactly as we do. Indeed, I have actually constructed such a thought experiment, to illustrate certain philosophical points about the separability of consciousness and behaviour.[5] But that is irrelevant to the actual causal role of consciousness in the real world.

When we are forming a thought experiment to test the evolutionary advantage of some phenotype, what are the rules of the game? In examining the evolutionary functions of wings, no one would think it allowable to argue that wings are useless because we can imagine birds flying just as well without wings. Why is it supposed to be allowable to argue that consciousness is useless because we can imagine humans and animals behaving just as they do now but without consciousness? As a science-fiction thought experiment, that is possible; but it is not an attempt to describe the actual world in which we live. In our world, the question 'What is the evolutionary function of consciousness?' is like the question 'What is the evolutionary function of being alive?' After all, we could imagine beings who outwardly behaved much as we do but are all made of cast iron and reproduce by smelting and who are all quite dead. I believe that the standard way in which the question is asked reveals fundamental confusions. In the case of consciousness the question 'What is the evolutionary advantage of consciousness?' is asked in a tone that reveals we are making the Cartesian mistake. We think of consciousness as not part of the ordinary physical world of wings and water, but as some mysterious non-physical phenomenon that stands outside the world of ordinary biological reality. If we think of consciousness biologically, and if we then try to take the question seriously, the question, 'what is the evolutionary function of consciousness?' boils down to, for example, 'what is the evolutionary function of being able to walk, run, sit, eat, think, see, hear, speak a language, reproduce, raise the young, organize social groups, find food, avoid danger, raise crops, and build shelters?', because for humans all of these activities, as well as a countless others essential for our survival, are *conscious activities*. That is, 'consciousness' does not name a separate phenomenon, isolable from all other aspects of life, but rather 'consciousness' names the mode in which humans and the higher animals conduct the major activities of their lives.

This is not to deny that there are interesting biological questions about the specific forms of our consciousness. For example, what evolutionary advantages, if any, do we derive from the fact that our colour discriminations are conscious and our digestive discriminations in the digestive tract are typically not conscious? But as a general challenge to the reality and efficacy of consciousness, the sceptical claim that consciousness serves no evolutionary function is without force.

Thesis 7

Causation is a relation between discrete events ordered in time. If it were really the case that brain processes cause conscious states, then conscious states would have to be separate events from brain processes and that result would be a form of dualism, dualism of brain, and consciousness. Any attempt to postulate a causal explanation of consciousness in terms of brain processes is necessarily dualistic and therefore incoherent. The correct scientific view is to see that consciousness is nothing but *patterns of neurone firings.*

Answer to thesis 7

This thesis expresses a common mistake about the nature of causation. Certainly there are many causal relations that fit this paradigm. So, for example, in the statement, 'the shooting caused the death of the man', we describe a sequence of events where first the man was shot and then he died. But there are lots of causal relations that are not discrete events but are permanent causal forces operating through time. Think of gravitational attraction. It is not the case that there is first gravitational attraction, and then, later on, the chairs and tables exert pressure against the floor. Rather, gravitational attraction is a constant operating force and, at least in these cases, the cause is contemporal with the effect.

More importantly for the present discussion, there are many forms of causal explanation that rely on bottom-up forms of causings. Two of my favourite examples are solidity and liquidity. This table is capable of resisting pressure and is not interpenetrated by solid objects. But, of course, the table, like other solid objects, consists entirely of clouds of molecules. Now, how is it possible that these clouds of molecules exhibit the causal properties of solidity? We have a theory: solidity is caused by the behaviour of molecules. Specifically, when the molecules move in vibratory movements within lattice structures, the object is solid. Now, somebody might say, 'Well, but then solidity consists in nothing but the behaviour of the molecules', and in a sense that has to be right. But solidity and liquidity are causal properties in addition to the summation of the molecule movements. Some philosophers find it useful to use the notion of an 'emergent property'. I don't find this a very clear notion, because it is so confused in the literature. But if we are careful, we can give a clear sense to the idea that consciousness, like solidity and liquidity, is an emergent property of the behaviour of the micro-elements of a system that is composed of those micro-elements. An emergent property, so defined, is a property that is explained by the behaviour of the micro-elements but cannot be deduced simply from the composition and the movements of the micro-elements. In my writings, I use the notion of a 'causally emergent' property[6] and, in that sense, liquidity, solidity, and consciousness are all causally emergent properties. They are emergent properties caused by the micro-elements of the system of which they are themselves features.

The point I am eager to insist on now is simply this: The fact that there is a causal relation between brain processes and conscious states does not imply a dualism of brain and consciousness any more than the fact that the causal relation between molecule movements and solidity implies a dualism of molecules and solidity. I believe the correct way to see the problem is to see that consciousness is a higher-level feature of the system, the behaviour of whose lower-level elements cause it to have that feature.

But this claim leads to the next problem—that of reductionism.

Thesis 8

Science is by its very nature reductionistic. A scientific account of consciousness must show that it is but an illusion in the same sense in which heat is an illusion. There is nothing to heat (of a gas), except the mean kinetic energy of the molecule movements: there is nothing else there. Now, similarly, a scientific account of consciousness will be reductionistic. It will show that there is nothing to consciousness except the behaviour of the neurones: there is nothing else there. And this is really the death blow to the idea that there will be a causal relation between the behaviour of the micro-elements, in this case neurones, and the conscious states of the system.

Answer to thesis 8

The concept of reduction is one of the most confused notions in science and philosophy. In the literature on the philosophy of science, I found at least half a dozen different concepts of reductionism. It seems to me that the notion has probably outlived its usefulness. What we want from science are general laws and causal explanations. Now, typically when we get a causal explanation, say of a disease, we can redefine the phenomenon in terms of the cause and so reduce the phenomenon to its cause. For example, instead of defining measles in terms of its symptoms, we redefine it in terms of the virus that causes the symptoms. So, measles is reduced to the presence of a certain kind of virus. There is no factual difference between saying, 'the virus causes the symptoms, which constitute the disease', and 'the presence of the virus just is the presence of the disease, and the disease causes the symptoms'. The facts are the same in both cases. The reduction is just a matter of different terminology. This is the point: what we want to know is, what are the facts?

In the case of reduction and causal explanations of the sort that I just gave, it seems to me that there are two sorts of reductions—those that eliminate the phenomenon being reduced by showing that there is really nothing there in addition to the features of the reducing phenomena, and those that do not eliminate the phenomenon but simply give a causal explanation of it. I don't suppose that this is a very precise distinction but some examples of it will make it intuitively clear. In the case of heat, we need to distinguish between the movement of the molecules with a certain kinetic energy on the one hand and the subjective sensations of heat on the other. There is nothing there except the molecules moving with a certain kinetic energy and this then causes in us the sensations that we call sensations of heat. The reductionist account of heat carves off the subjective sensations and defines heat as the kinetic energy of the molecular movements. We have an eliminative reduction of heat because there is no objective phenomenon there except the kinetic energy of the molecular movements. Analogous remarks can be made about colour. There is nothing there but the differential scattering of light and these cause in us 'colour experiences'. But there isn't any colour phenomenon there beyond the causes in the form of light reflectances and their subjective effects on us. In such cases, we can do an eliminative reduction of heat and colour. We can say there is nothing there but the physical causes and these cause the subjective experiences. Such reductions are eliminative reductions in the sense that they get rid of the phenomenon that is being reduced. But in this respect they differ from the reductions of solidity to the vibratory movement of molecules in lattice structures. Solidity is a causal property of the system that cannot be eliminated by the reduction of solidity to the vibratory movements of molecules in lattice-type structures.

But now why can't we do an eliminative reduction of consciousness in the way that we did for heat and colour. The pattern of the facts is parallel. For heat and colour we have physical causes and subjective experiences; for consciousness we have physical causes in the form of brain processes and the subjective experience of consciousness. So it seems we should reduce consciousness to brain processes. And, of course, we could if we wanted to, at least in this trivial sense. We could redefine the word 'consciousness' to mean the neurobiological causes of our subjective experiences. But if we did, we would still have the subjective experiences left over, and the whole point of having the concept of consciousness was to have a word to name those subjective experiences. The other reductions were based on carving off the subjective experience of heat, colour, etc., and redefining the notion in terms of the causes of those experiences. But where the phenomenon that we are discussing is the subjective experience itself, you cannot carve off the subjective experience and redefine the notion in terms of its causes, without losing the whole point of having the concept in the first place. The asymmetry between heat and colour on the one hand and consciousness on the other has not to do with the facts in the world, but rather with our definitional practices. We need a word to refer to ontologically subjective phenomena of awareness or sentience. And we would lose that feature of the concept of consciousness if we were to redefine the word in terms of the causes of our experiences.

You can't make the appearance–reality distinction for conscious states themselves, as you can for heat and colour, because, for conscious states, the existence of the appearance is the reality in question. If it seems to me I am conscious then I am conscious. And that is not an epistemic point. It does not imply that we have certain knowledge of the nature of our own conscious states. On the contrary, we are frequently mistaken about our own conscious states—as, for example, in the case of phantom limb pains. It is a point about the ontology of conscious states.

When we study consciousness scientifically, I believe we should forget about our old obsession with reductionism and seek causal explanations. What we want is a causal explanation of how brain processes cause our conscious experiences. The obsession with reductionism is a hangover from an earlier phase in the development of scientific knowledge.

Thesis 9

Any genuinely scientific account of consciousness must be an information-processing account. That is, we must see consciousness as consisting of a series of information processes, and the standard apparatus that we have for accounting for information processing in terms of symbol manipulation by a computing device must form the basis of any scientific account of consciousness.

Answer to thesis 9

I have already answered this mistake in detail elsewhere.[7] For present purposes, the essential thing to remember is this: consciousness is an intrinsic feature of certain human and animal nervous systems. The problem with the concept of 'information processing' is that information processing is typically in the mind of an observer. For example, we treat a computer as a bearer and processor of information but, intrinsically, the computer is simply an electronic circuit. We design, build, and use such circuits because we can interpret their inputs, outputs, and intermediate processes as information bearing. But in such a case the information in the computer is in the eye of the beholder, it is not intrinsic to the computational system. What goes for the concept of information goes *a fortiori* for the concept of 'symbol manipulation'. The electrical-state transitions of a computer are symbol manipulations only relative to the attachment of a symbolic interpretation by some designer, programmer, or user. The reason we cannot analyse consciousness in terms of information processing and symbol manipulation is that consciousness is intrinsic to the biology of nervous systems; information processing and symbol manipulation are observer-relative.

For this reason, any system at all can be interpreted as an information-processing system. The stomach processes information about digestion; the falling body processes information about time, distance, and gravity. And so on.

The exception to the claim that information processing is observer-relative are precisely cases where some conscious agent is thinking. If I as a conscious agent think, consciously or unconsciously, '2 + 2 = 4', then the information processing and symbol manipulation are intrinsic to my mental processes, because they are the processes of a conscious agent. But in that respect my mental processes differ from my pocket calculator adding 2 + 2 and getting 4. The addition in the calculator is not intrinsic to the circuit, the addition in me is intrinsic to my mental life.

The result of these observations is that to make the distinction between the cases which are intrinsically information bearing and symbol manipulating from those which are observer-relative we need the notion of consciousness. Therefore, we cannot explain the notion of consciousness in terms of information processing and symbol manipulations.

Conclusion

There are other mistakes I could have discussed, but I hope the removal of those I listed will help us make progress in the study of consciousness. My main message is that we need to take consciousness seriously as a biological phenomenon. Conscious states are caused by neuronal processes, they are realized in neuronal systems, and they are intrinsically inner, subjective states of awareness or sentience.

We want to know how they are caused by and realized in the brain. Perhaps they can also be caused by some sort of chemistry different from brains altogether, but until we know how brains do it we are not likely to be able to produce it artificially in other chemical systems. The mistakes to avoid are those of changing the subject—thinking that consciousness is a matter of information processing or behaviour, for example—or not taking consciousness seriously on its own terms. Perhaps, above all, we need to forget about the history of science, and get on with producing what may turn out to be a new phase in that history.

Notes and references

1. See, for example, J. R. Searle, *Minds, brains and science* (Harvard University Press, Cambridge, Mass., 1984) and J. R. Searle, *The rediscovery of the mind* (MIT Press, Cambridge, Mass., 1992).
2. René Descartes, *Meditations on first philosophy*, especially Third Meditation; for example, 'But in order for a given idea to contain such and such objective reality, it must surely derive it from some cause which contains at least as much formal reality as there is objective reality in the idea.' (*The philosophical writings of Descartes*, Vol. 2, trans. J. Cottingham, R. Stoothoff, and D. Murdoch (Cambridge University Press, Cambridge, 1984).)
3. See, for example, his article, 'What is it like to be a bat?' in *The Philosophical Review*, **83**, 434–50 (1974).
4. For example, Francis Crick, *The astonishing hypothesis: the scientific search for the soul* (Simon and Schuster, New York, 1994) and Gerald Edelman, *The remembered present: a biological theory of consciousness* (Basic Books, New York, 1989).
5. Searle, *The rediscovery of the mind*, Ch. 3.
6. See Searle, *The rediscovery of the mind*, Ch. 5, pp. 111ff.
7. See J. R. Searle, 'Minds, brains, and programs'. *Behavioral and Brain Sciences*, **3** (1980), pp. 415–57. See also Searle, *Minds, brains and science* and *The rediscovery of the mind*.

....

CHAPTER 3

Biological variability and brain function

OLAF SPORNS

Variability and population thinking

The first two chapters of Darwin's *Origin of species* contain an account of variation under conditions of domestication and under nature. Darwin's own words may still serve as a clear exposition of the problem of biological variability:

> The many slight differences which appear in the offspring from the same parents ... may be called individual differences. No one supposes that all the individual of the same species are cast in the same mould. The individual differences are of the highest importance to us, for they are often inherited, as must be familiar to every one; and they thus afford materials for natural selection to act on and accumulate, in the same manner as man accumulates in any given direction individual differences in his domesticated productions. ... I am convinced that the most experienced naturalist would be surprised at the number of the cases of variability, even in important parts of structure, which he could collect on good authority, as I have collected, during a course of years. It should be remembered that systematicists are far from being pleased at finding variability in important characters, and that there are not many men who will laboriously examine internal and important organs, and compare them in many specimens of the same species. It would never have been expected that the branchings of the main nerves close to the great central ganglion of an insect would have been variable in the same species; .. yet Sir J. Lubbock has shown a degree of variability in these main nerves in Coccus, which may almost be compared to the irregular branching of a stem of a tree. (Darwin 1859/1962, pp. 59–60.)

It was not accidental that the *Origin* started with a discussion of biological variability. Darwin's solution to the 'species question', the theory of evolution by natural selection, reserves a special place for variation within populations of individuals. According to Darwin, 'species are only strongly marked and permanent varieties, and ... each species first existed as a variety'. His realization that biological variability is absolutely essential for the emergence and continuing evolution of new species had the most profound impact on biological thought. Evolutionary theory now regards variation as one of the fundamental requirements for natural selection. Darwin's theory introduced a new way of thinking into biology, called 'population thinking' by the evolutionary biologist Ernst Mayr (Mayr 1959, 1963; see also Dobzhansky 1967).

> The assumptions of population thinking are diametrically opposed to those of the typologist. The populationist stresses the uniqueness of everything in the organic world. ... All organisms and organic phenomena are composed of unique features and can be described collectively only in statistical terms. ... The ultimate conclusions of the population thinker and of the typologist are precisely the opposite. For the typologist, the type (*eidos*) is real and the variation an illusion, while for the populationist the type (average) is an abstraction and only the variation is real. No two ways of looking at nature could be more different. (Mayr 1959, p. 2.)

Direct observation of numerous animal species as well as humans convinced Darwin that behaviour as well as higher mental phenomena have evolutionary histories of their own, just as the muscles and bones of the animals exhibiting them. Occasionally (as in the quote above), Darwin refers to the variable nature of brain anatomy, a reflection of his insight that the brain, just as any other part of an organism, is the product of evolution and thus subject to variation. In the first half of the nineteenth century the scientific study of the brain as the seat of mental phenomena and the originator of behaviour was in its infancy, and it was up to later generations of scientists to link the natural history of the brain to the evolution of behaviour and cognition. Meanwhile, the issue of neuronal variability became submerged outside of a purely evolutionary context. In fact, very little has been written on neuronal variability *per se*, and almost nothing on the relationship of neuronal variability to function, or on the mechanisms by which neuronal variability might be generated.

Among the few scientists to discuss the problem was the neurophysiologist Karl Lashley. In an article containing findings from an anatomical comparison of several postmortem brains he concluded:

> The brain is extremely variable in every character that has been subjected to measurement. Its diversities of structure within the species are of the same general character as are the differences between related species or even between orders of animals. ... Individuals start life with brains differing enormously in structure; unlike in number, size, and arrangement of neurons as well as in grosser features.

> The variation in cells and tracts must have functional significance. (Lashley 1947, p. 333.)

Lashley suspected that many (if not most) variations are genetic in origin and proposed a link between variants at the extreme of the normal distribution and the appearance of pathological symptoms. Thus, Lashley appears to have anticipated the potential influence of anatomical variability on individual behaviour, although he did not see the broader implications of this variability for neural function and said nothing about its origins (except that they are possibly hereditary).

Within the last 20 to 30 years modern neuroscience has provided more knowledge about all aspects of brain function than in all of previous history combined. Much of neuroscience research has been carried out in attempts to characterize the unvarying species-specific aspects of brain anatomy and physiology while stressing the causal and deterministic actions of the brain's components. On the whole, although with some notable exceptions, population thinking and the emphasis on biological variability have had little direct impact on neuroscience. The problem of variability, however, so trenchantly analysed by Darwin and of recognized importance in evolutionary theory, remains a challenge to neuroscience.

In this chapter I present evidence concerning variability within the nervous system and ask about the consequences of these observations for theoretical accounts of the brain. As we will see, the evidence overwhelmingly supports the view that the brain is as variable in its structural arrangement and emerging functionality as any other part of an organism. This variability persists throughout the lifetime of the organism and forms an important basis for selectional processes acting within the nervous system in somatic time (Edelman 1987). Variability is a unique property of all organisms, and it sets them apart from inanimate objects and phenomena that are within the realm of the physical sciences. My conclusion is that the individual history of an organism with its lack of closed causal explanations for most of the detail and some global patterns of its morphological and behavioural phenotype, as well as its cognitive and perceptual capabilities, refutes all attempts to reduce biology to special manifestations of universal (physical) laws.

Sources of biological variability

Before examining in some detail to what extent biological nervous systems vary we need to briefly discuss the main sources of variability in organisms. Differences in the phenotype of individual organisms can come about because of differences in developmental or other epigenetic processes unfolding over time. Much of the variation in phenotypic appearance found in natural populations (including the examples discussed by Darwin) can be accounted for by genetic differences among individuals.[1] The second major source of variability is epigenetic (literally 'after' or 'over' the genes). On their way from the fertilized egg to the adult specimen all animal organisms undergo profound phenotypic changes in the course of highly complex developmental processes following a macroscopically predictable and constant pattern. The precise course of events at the cellular and molecular level that gives rise to a living organism, however, is never quite the same between individuals. First, pre-existing genetic differences can alter the course of developmental events and lead to variant phenotypes. In this instance, genetic and epigenetic variability appears inextricably linked. Second, even if genetic differences are negligible or can be ruled out (as in the case of monozygotic twins) the exact historical sequence of developmental events will tend to differ and may result in individuals that differ in their phenotype. These events are under the control of developmental regulatory processes, such as those involved in the adhesion, movement, differentiation, growth, division, and death of cells.

The generation of diversity in growing neuronal structures is a necessary outcome of the action of morphoregulatory controls on cellular driving forces that are sensitive to local influence and context during development (Edelman 1988). An absolutely central point for our discussion of variability in the nervous system is that epigenetic processes act continuously throughout the lifetime of an organism, even after embryonic development has ceased and the adult body structure and function have emerged (Edelman 1987). The adult nervous system becomes the staging ground for neuronal epigenetic processes that leave traces reflecting as well as shaping our experience of the world, some slight and ineffectual, others significant and of almost instantaneous consequence for the organism. Major instruments in this ongoing process are, among many others, synaptic changes accompanying neuronal activity, the re-allocation of neurones to varying tasks, and the reorganization and regeneration of neuronal tissue in response to major changes in sensory input or following injury. All of these will be discussed in some detail below. For now, the essential point is that no two individuals of the same species, no matter how close their actual environments and ecological conditions may be, will ever experience exactly the same sequence of neuronal epigenetic processes.

The degree to which the neuronal morphology and behavioural phenotype of individuals will differ will depend on a variety of factors, such as on actual differences in environment, as well as on constraints of adaptation. But differ they must, as their nervous systems continue to embody and reflect their unique experiences. Neuronal variability, due to a combination of genetic and epigenetic processes among different individuals, or due to epigenetic processes alone within a single brain, becomes almost a default condition (Edelman 1992). Another important point is that in cases where biological organisms belonging to the same species differ in their morphological or behavioural phenotype, it can be very difficult to attribute this difference to a single cause, such as a gene or a particular epigenetic process or event. Genetic and epigenetic processes are often operating in parallel and, consequently, their effects on variability often combine and become inseparable. The fact that even genetically identical organisms can show significant variation in neuronal phenotypic traits, however, gives some ground for the assumption that epigenetic processes alone can (and generally will) generate phenotypic variants.

Evidence for variability in the nervous system

Variability in the nervous system can be expressed in either structure or function. As we consider these aspects in turn, it will become clear that this distinction is somewhat arbitrary as structural and functional variability often occur together.

Structural variability

Within the nervous system, structural variability refers to the existence of morphological, physiological, or biochemical differences in brain structure between individuals. If we consider structures that are repeated over and over within the nervous system and subserve identical function (as in the case of single neurones of a given type or class, cortical columns, or modular functional circuits) we may also consider the variability of these structures within one individual organism.[2] Examples for structural variability are differences in the size or location of brain areas or the variations in the branching patterns of single neurones. As a first approximation, all of the instances of structural variability can be examined by comparing different individuals or repeated structures without consideration of time.

The structural variability of the brain does not usually affect gross anatomical features, which are characteristic for the animal species, although it has been known for a long time that the pattern of gyri and sulci on the surface of the human cortex is highly variable between individuals. Modern methods have shown that the size and relative position of specific brain areas can vary between individuals on a scale of millimetres or even centimetres. Functional neuroimaging of the human brain has revealed that the positioning of cortical areas associated with a variety of cognitive or behavioural tasks shows significant interindividual variation. For example, a cortical region of the human brain responding to moving stimuli (a presumed analogue of the non-human primate area V5) and functionally mapped by positron emission tomography is located in the lateral occipital lobe, but its exact position can differ by as much as 2.7 centimetres in different subjects (Watson *et al.* 1993). In a recent anatomical study of cortical areas in the human frontal lobe, Rajkowska and Goldman-Rakic (1995) reported significant morphological variability in their location and size in the brains of seven individuals.

As these and other findings indicate, both the functional and anatomical localization of cortical areas differ significantly between individuals. Researchers mapping the human brain are arriving at the conclusion that classical maps of cortical areas can at best represent group averages[3] and can be inadequate when applied to imaging data obtained from an individual brain. Measurements of the surface area of the brains of monozygotic twins suggest that interindividual variability is reduced for genetically identical pairs, while significant and hemisphere-specific variability exists for genetically different individuals (Tramo *et al.* 1995). Apparently, genetic factors, possibly involved in the control of morphoregulatory processes during development, contribute a major component to the variability of macroscopic brain structure in adults.

The microanatomy of neurones and neuronal circuits shows significant variation, both between individuals that vary in their genetic background, between individuals of identical genetic background and between thousands (and perhaps millions) of neurones and circuits of a given type repeated within a single individual brain. This appears to be true for all animal species that have been investigated, including species with smaller nervous systems containing relatively few and often highly specialized neurones. Some structural parameters such as overall branching pattern and target connections are relatively constant among homologous neurones in different individuals of some invertebrate species (see examples in Purves 1988), but other cases show remarkable variation. The sensory neurones in the femoral tactile spine of the cockroach exhibits considerable variability in the size and shape of the neuronal cell body, while the morphology of the dendrite and axon are more consistent (French *et al.* 1993). In contrast, the branching pattern of the descending contralateral movement detector in the meta-thoracic ganglion of the locust is highly variant (Pearson and Goodman 1979; for other examples see Edelman 1987).

In general, even in cases where single neurones are dedicated to performing specific functions (and structural variation can therefore not be absorbed by redundancy) these neurones sometimes show dramatic variation. In vertebrates homologous neurones almost always differ in fine morphology from one individual to another, or, in the case of repeated structures, within an individual. For example, no two pyramidal cells in the cerebral cortex of a vertebrate are alike in the number and pattern of their incoming and outgoing synaptic connectivity. A scientist shrunk to about the size of a hundredth of a millimetre and transported into the jungle of nerve cells and connections at homologous places in the brains of two individuals would report an overall similarity in the types of cells and some statistical aspects of the connectivity, but he would be unable to trace any single connection or map a single neurone's inputs and outputs in one individual and obtain a perfect match to the other. To make matters worse, if he were to move laterally within a given anatomical region of the brain, the shape and branchings of axons and dendrites, the positions of the cell bodies, and the spatial arrangements of synaptic endings would differ from place to place.[4]

Genetic control of anatomical structures at this level of scale can be excluded due to the vast number of neurones and connections that have to be accounted for. Variable neuroanatomy is the inevitable outcome of the action of multiple epigenetic mechanisms governing the motion, adhesion, differentiation, and death of cells and the outgrowth, target selection, and consolidation of their axonal and dendritic processes (Edelman 1988). Variability ensues because the complex sequence of events involving uncountable molecular and cellular interactions that eventually give rise to a nervous system will necessarily differ between individuals.

Further examples of variability include biochemical and physiological parameters determining individual synaptic responses (Sporns and Jenkinson 1997), the distribution of neurotransmitters and peptides (Mai *et al.* 1993), as well as receptor densities (Farde *et al.* 1995) throughout the human central nervous system. This interindividual variability may have implications for individual differences in mental function and disease. How much of biochemical variability is due to epigenetic mechanisms, and how much to genetic factors, is unknown.

Functional variability
Functional variability can be distinguished from structural variability as it refers to differences in functional processes occurring in time. As with structural variability, these differences may exist between individuals or, if processes are repeated sequentially, they may exist within one individual. Examples are provided by the variable nature of bodily movements or entire behavioural sequences. A point to be stressed is that structural variability in the brain often implies functional variability. The co-operative

action of functional neuronal elements that show structural variation gives rise to dynamics that is itself highly variable (within a broad envelope of constancy).

A key example of developmental plasticity and behavioural changes depending upon actual experience is the development of motor behaviour in many animal species. Studies of the early postnatal development of human neonates (Thelen and Corbetta 1994; Thelen and Smith 1994) demonstrate that they are not born with a fixed and hard-wired set of movements; instead the ability to move—for example, reach for and grasp objects—develops over time. Individuals differ significantly in their progress during learning and in their acquired movement patterns. Such variability also exists in other vertebrate species, in some cases even before birth. Chicks at different embryonic stages show spontaneous hindlimb movements that are precursors of later postnatal behaviours (Bradley and Bekoff 1990). These movements conform to a global pattern but from instance to instance different muscle groups and biomechanical linkages are engaged. Swimming movements in newborn rats are also characterized by highly variable temporal sequences and grouping of muscles (Cazalets et al. 1990). Several muscles in the hindlimb of the cat show significant levels of interanimal variability in activity during stereotyped movements or reflexes even in the adult (Loeb 1993). These few examples, all taken from movement physiology, illustrate the point that the emergence of motor skills in vertebrates is generally accompanied by an extended period of relative instability and variability, which is gradually reduced as the skill evolves but never quite eliminated.

Variability and continuous remodelling of structure and function

In the nervous system of a living organism, structural and functional variabilities are not strictly separable. They appear to be intricately linked in processes that occur over time and leave structural traces. Many different aspects of neuroplasticity, including structural alterations of neuronal circuitry and changes of synaptic efficacy, contribute to ongoing structural and functional remodelling. The important point in the context of this discussion is that activity- or experience-dependent plasticity can lead to variability both within and between individual organisms as it reflects their respective histories and replenishes the reservoir of variability available to somatic selection. Plasticity, in many cases, is responsible for the successful adaptation of an organism to an environmental context and in that sense produces structural and functional alterations that are non-random and seemingly purposeful. Adaptational biases, however, create only an envelope of possibilities, especially at the microscopic level of synaptic patterning; individual solutions found by an organism in response to an environmental challenge will differ at least in detail, if not in global pattern. Thus, adaptation does not contradict continuing variation; in a selectional system, the former in fact requires the latter.

In some sense, ongoing variability can be regarded as the continuation into adult life of basic developmental processes. For example, in all vertebrate species that have been examined, the formation and elaboration of neuronal circuitry is far from finished at the end of embryogenesis. Postnatal developmental processes continue to shape neuroanatomy, continuing into adulthood. Synaptogenesis, the formation of synaptic contacts between neurones, is a useful indicator of the extent of neural development, particularly the elaboration of neural circuitry. Studies in several vertebrate species (see, for example, Blue and Parnavelas 1983), including humans (Huttenlocher 1979), indicate that synaptic densities (the number of synapses per unit volume of brain) continue to increase throughout embryogenesis and early postnatal life, and decline thereafter. Rakic and coworkers measured the synaptic density in the prefrontal cortex of the macaque (Bourgeois and Rakic 1993) and rhesus monkey (Bourgeois *et al.* 1994) over the entire lifespan of the organism. In the rhesus monkey, an early rapid phase of synaptogenesis (between 2 months prenatal to 2 months postnatal) was followed by a plateau period (lasting until 3 years) and a progressive decline thereafter. In the macaque monkey, this decline amounts to the loss of about 5000 synapses per second in the visual cortex alone between 2.7 and 5 years (Bourgeois and Rakic 1993).

Even though the overall synaptic density does not take into account synaptic turnover and the ongoing remodelling of cortical circuits (Cotman and Nieto-Sampedro 1984; Bailey and Kandel 1993) these measurements give an impression of the major structural fluctuations that the cortical system undergoes in somatic time. It is unknown what determines which synapses disappear when. There is no sign of an instructive signal delivering death warrants to individual cells or synapses. The degree to which the microanatomy of the brain appears to fluctuate (Greenough and Chang 1988; Wolff *et al.* 1995) poses a major problem for the overall stability and constancy of perceptual and cognitive abilities, including memory. It is obvious in the light of the evidence that a population-based account of cortical function seems more likely to succeed than others that use single cells as functional units and that require precise point-to-point writing.

The experience-dependent formation of intracortical circuitry has been studied extensively in the primary visual cortex of several species. The system of tangential connections in area 17 of the cat develops mainly after birth (see, for example, Callaway and Katz 1991). Selectivity of the connection pattern (preferential connectivity between groups of neurones that have similar response properties) is achieved after pruning (elimination) of inappropriate connections. Löwel and Singer (1992) reported that the criterion for the selection of tangential connections during visual experience is correlated neural activity.

Such use-dependent selection based on correlations provides a general mechanism for the generation of functional anatomical connectivity throughout the cortex. Succinctly stated, 'neurones wire together if they fire together'. Other studies by Singer and colleagues suggest that while temporal correlation between neuronal groups serves selectively to stabilize cortical circuitry, the same correlations play an important role (after the connectivity has been laid down) in several visual functions (Singer and Gray 1995). Differences in visual experience can lead to radical or more gradual alterations in the patterning of intracortical connections.

The assignment of cortical surface area to different functional or sensory modalities is not fixed during the lifetime of an organism. Early studies by Merzenich and his colleagues demonstrated differences in the size and location of somatosensory cortical areas in the macaque monkey (Merzenich et al. 1987). Experimental manipulations of the input to the somatosensory cortex were able to induce rapid and long-lasting changes in the parcellation of the cortical surface. Merzenich et al. (1984) found that after amputation of a digit of the hand the cortical region previously responsive to that digit does not turn into a 'blank' non-responsive area, but becomes responsive to neighbouring digits. This was originally observed within 2 months of amputation. Extensive stimulation of one digit results in an enlarged representation of this digit on the cortical surface, while conjoint stimulation of two adjacent digits results in neurones whose receptive fields include parts of both digits. The time-course of reorganization appears to be relatively fast in almost all cases; in experiments similar to those of Merzenich et al., Calford and Tweedale (1990) observe effects within 20 minutes. A few hours of intracortical microstimulation were sufficient to induce changes in the pattern of cortical representation (Recanzone et al. 1992; Dinse et al. 1993), leading to enlargement as well as shifting of areal and modality borders (Spengler and Dinse 1992). Correlated neural activity is crucial for these effects, pointing to the importance of functionally coupled neuronal groups in the formation and maintenance of cortical representations.

More recently, Pons et al. (1991) found that cortical reorganization can extend over several millimetres, up to 1 centimetre and more of cortical surface. Several years after amputation of an upper limb the cortical area corresponding to that limb becomes responsive to sensory inputs from the face. Ramachandran (1993) has investigated the perceptual correlates of this cortical reorganization in humans and found that a perceptual response consistent with cortical reorganization can be obtained as early as 4 weeks after loss of an arm. A different example, illustrating cortical reorganization accompanying 'normal' activity, comes from a study of Elbert *et al.* (1995). A comparison of individuals experienced in string playing versus non-experienced individuals found that the amount of cortical surface allocated to the left hand of string players was increased.

All these examples show that changes in the afferent input to the cortex can result in reorganization of functional cortical anatomy. The evidence suggests that cortical reorganization is an ongoing process occurring throughout the lifetime of an organism, although its range and degree may decrease somewhat with increasing age of the organism. As cortical reorganization reflects the individual's sensory and motor experience, it contributes to increased neuronal variation between individuals (see also Almassy *et al.* 1998).

Experience-dependent variations in important morphological parameters occur even in invertebrates, who were regarded (at least among 'cortical chauvinists') more as hard-wired robots than as highly adaptive and plastic organisms. Heisenberg *et al.* (1995), in a detailed study of variations in brain morphology of *Drosophila*, report that structural plasticity is present in most parts of the fly's brain and at various times during its life cycle. This plasticity seems to be related to the fly's living space and social environment. The result is populations of flies highly variable in the respective volume of several important brain regions. The authors conclude that 'Whatever the particular behavioral correlates of the observed volume changes may be, our investigation suggests that not only in vertebrates but also in arthropods sprouting and decay of terminal branches as well as the concomitant turnover and reorganization of synapses occur continuously and in many, if not all, regions of the brain.' Heisenberg's observations argue for experience-dependent epigenetic processes modulating fine and gross anatomical features in an organism in which (according to popular notions) all important aspects of function (including those of the nervous system) are under tight genetic control. The days of the 'hard-wired' insect may be numbered.

Variability and brain theory

Taken together, the evidence from numerous studies involving both vertebrate and invertebrate species argues for a significant degree of biological variability within the nervous system. What is the relationship of this variability to neuronal function? Following Darwin, Mayr and evolutionary theory, neuronal variability of structure and function, generated by genetic and epigenetic factors, is of central importance in theoretical frameworks based on population thinking and selection. Surprisingly, the contrasting notions of typological thinking and instructionism have traditionally been very influential in brain theory. I will briefly examine the contrast between these notions and examine their relationship to the issue of variability.

Instructionist theories in biology stress the transfer of information from one domain (such as the environment) to another (such as a biological system). The biological system will—by means that differ from case to case—embody or store the information it has received. The system's construction and ongoing operation depend crucially on the transfer of this information from its environment.[5] Selectionist theories, on the other hand, generally do not require such information transfer in their construction. Rather they work on the basis of three principles: (1) pre-existing variation among components within the biological system; (2) encounter of the biological system with an environment and polling of the system's components; and (3) differential amplification of those components of the system that meet a threshold criterion after encounter. (The criterion in evolutionary selection is fitness.) Such a system, while not containing any structures or fixed rules that are directly transferred from the environment, will behave *ex post facto* as if the system has been instructed. This is a crucial point: after the fact, systems that work according to instruction or selection may appear to perform alike, but acquiring and maintaining this performance are achieved by very different principles and mechanisms in both cases.

One of the most important distinctions between the two approaches is the relevance of variation among components of the biological system. Because it is based on information transfer, instructionism does not require any pre-existing structure within the system (except insofar as it relates to the successful transfer and storage of information): it can do with a *tabula rasa.* Variation is simply 'noise' and has no functional importance *per se* except to degrade performance. A selective system, on the other hand, has a repertoire of pre-existing variable functional components, in addition to ways of differentially amplifying those that perform best or match to the environment. In this case, variation is centrally important in achieving good adaptive performance of the system. While variability may be structural or functional, structural differences between functional elements can give rise to additional variability in the ensuing dynamics. In summary, the tremendous variability in nervous systems acquires functional meaning only if viewed as a basis for selectional mechanisms.

Adopting a selectionist perspective changes our popular notion of the brain as an information-processing device. Throughout modern history the prevailing popular view of the brain has been influenced by metaphors borrowed from some of the most advanced technology of the time. Descartes compared nerves to the systems of waterpipes that were used in animated figures. According to Descartes, animals were merely machines; all their actions resulted from mechanical processes and were governed by natural (which means physical) laws. Humans differed in that they had an immaterial soul that was not subject to such laws and, in ways unexplained, interacted with the body. The mechanistic and dualistic viewpoint underlying Descartes's vision provided a tremendous boost to the development of science and remains to this day perhaps the most influential view of how the brain and the mind work. The building of automata that would replicate human or animal behaviour has a long tradition (reviewed in Reeke and Sporns 1993) that is essentially rooted in Cartesian dualism.

Most recently, the most powerful metaphor to understand brain function and the functions and properties of (human) minds has certainly been the computer. The computer metaphor basically implies that the brain is a piece of specialized hardware, and the mind a kind of software implemented to run on this hardware (Fodor 1983; Pylyshyn 1984; Johnson-Laird 1988). This view forms part of a philosophy of mind called functionalism, which essentially postulates that mental phenomena are rule- or instruction-based and computational in nature. As such they are independent of particular material implementations. It should be noted that the use of the computer as an explanatory model for brain function and the underlying philosophy of functionalism have been subject to severe criticism both from neuroscientists and from philosophers of mind (Putnam 1988; Edelman 1992; Searle 1992). Considered from a biological perspective, functionalism runs counter to the general recognition of the importance of evolutionary and developmental constraints on morphology (Reeke and Sporns 1990). The high degree of variability within the nervous system is irreconcilable with theories based on the manipulation of information (such as functionalism); machines specialized for such functions (like computers) usually require precise wiring at all levels and scales. Indeed, most information-processing theories of the brain perpetuate the erroneous conclusion that complex behaviours must be carried out by precise circuitry.

Variability and brain modelling

An increasing amount of experimental evidence supports selectionist views of the brain, coming from developmental biology, neurophysiology, and cognitive science (summarized in Edelman 1993; Sporns and Tononi 1994).

Another way to show the self-consistency of selectionism is to construct detailed computer models of neuronal networks and allow them to interact through sensory and motor organs with an environment containing objects and events. This approach, called synthetic neural modelling (Reeke *et al.* 1990), has been applied to a variety of problems, including perceptual categorization, the acquisition of motor skills, autonomous behaviour, and learning. Variability is a crucial component of each model's design and operation. In most cases, connectivity patterns—globally defined as linkages between identified neuronal regions—are generated with local random variation. This has several important consequences for the computer model (Reeke *et al.* 1990; Edelman *et al.* 1992; Almassy *et al.*):

1. Internal firing patterns, even in response to the same stimulus presented in the environment, will differ between different presentations.
2. If two individuals of the same overall design are 'initialized' with different random variations in their neuroanatomy, their behaviour and future developmental and experiential histories will differ.
3. If computer models are embedded in a real-world artefact, no two 'runs' of the same model (even if started as exact copics including all random variations) will come out identical. The course of stimulation and behaviour will inevitably differ and so will the resulting experience-dependent plastic changes accruing in the model's networks.

The use of variability in a selectional model is perhaps best illustrated in the following example, related to studies of the early development of reaching movements in infants (see above). Our modelling studies have shown that variable actions can form a repertoire from which adaptive or value-linked behaviours (and their corresponding neural elements) can be selected (Reeke *et al.* 1990). This forms the basis for a recent proposal to address the problem of motor development and the acquisition of motor skills in the context of selection (Sporns and Edelman 1993). The progression from a set of simple 'synergies' to well-adapted movements can be viewed as a process of selection of initially variable movements from a repertoire. During ongoing experience and interaction with the environment appropriate movements are selected and others are eliminated, in response to global value signals related to the salience of these movements to the organism. This selectionist view of motor development avoids typical pitfalls of proposals based on instruction. It does not require the system to find an explicit solution to the computationally hard problem of inverse kinematics (finding a set of motor commands in a multidimensional redundant space for a given three-dimensional trajectory). Thus the problem of simultaneously controlling many degrees of freedom

(Bernstein 1967), which presents insurmountable problems to an instructionist system based on computation and fixed rules, is solved naturally by selection of variant movements from a repertoire.

While these computational studies are necessarily incomplete and do not ultimately prove the validity of selectionist thinking as applied to the brain, they serve as a valuable testing ground for the effects of variability on global neuronal function. In addition, they serve as illustrations of the magnitude of variation in both structure and function even in simple nervous systems and their relationship to their perceptual and behavioural capabilities.

Variability and the origin of individuality

If biological variability is central, reductionist explanations of brain function must fail, because they do not account for its origins. In Gerald Edelman's words 'To reduce a theory of an individual's behavior to a theory of molecular interactions is simply silly' (Edelman 1995, p. 201). In a recent article the renowned physicist Steven Weinberg takes issue with several critiques of reductionism. In particular, he disagrees with the viewpoint that biological processes and phenomena such as mind and life cannot be explained by the fundamental laws of physics alone. 'The rules they obey are not independent truths, but follow from scientific principles at a deeper level; *apart from historical accidents* that by definition cannot be explained' (Weinberg 1995, p. 40). He goes on to argue that selectionist theories of the mind do not rule out thoroughly reductionist views of mentality. Countering Edelman's rejection of 'silly reductionism' (Edelman 1995) he states that

> It may or may not be silly to pursue reductionist programs of research on complicated systems that are strongly conditioned by history, like brains or roses or thunderstorms. What is never silly is the perspective, provided by reductionism, that *apart from historical accidents* these things ultimately are the way they are because of the fundamental principles of physics. (Weinberg 1995, p. 41; my italics.)

Weinberg's plea for reductionism in biology brushes aside that which cannot be reduced—the accidents of individual history.[6] Biological organisms are unique, however, in the degree to which they vary within populations and in their use of variation as a fundamental basis for the principle of selection. An essential component in generating biological variability is the uniqueness in the series of events that gives rise to organisms, over the time-scale of evolution as well as over the lifetime of an individual. The variability of the nervous system—structural, dynamic, and ongoing—poses a challenge to any biologically based theoretical account of brain function. If we begin to consider neuronal variability in the context of selection as an active principle in brain function we will open up a new view of the brain as the carrier of individuality. Edelman's theory of neuronal group selection (Edelman 1987, 1989), a comprehensive account in selectionist terms of virtually all aspects of neuronal function, from neural development to consciousness, provides the only biologically based brain theory that deals with the issue of human individuality head-on:

> The flux of categorization ... is an individual and irreversible one. It is a history. ... Given the diversity of the repertoires of the brain, it is extremely unlikely that any two selective events, even apparently identical ones, would have identical consequences. Each individual is not only subject, like all material systems, to the second law of thermodynamics, but also to a multi-layered set of irreversible selectional events in his or her perception and memory. Indeed, selective systems are by their nature irreversible. (Edelman 1995, p. 203.)

The future focus of brain science as it relates to human identity will be on that which is irreversible and irreducible. No two brains are alike, because, in organisms, history does not repeat itself.

Acknowledgement

The work performed at The Neurosciences Institute cited here was supported by the Neurosciences Research Foundation.

Notes

1. Darwin himself as well as his contemporaries had no knowledge of the sources of variation. They were therefore also unaware of the distinction between genetic and epigenetic factors (a distinction approximated by heritable and acquired characters).
2. Note that the genetic component of variability is therefore nil.
3. The best known of these maps drawn in 1909 by Korbinian Brodmann is based on a single case.
4. And to add to the confusion, if the host animal carrying the shrunken scientist is alive and its brain active, neural morphology at a given locale would actually undergo plastic changes within hours, perhaps minutes (see later).
5. For a recent controversy on this point see the article by Quartz and Sejnowski (1997) and my commentary (Sporns 1997).
6. Although not explicitly discussed, Weinberg would have to include the accidents of evolutionary history as well, from mass extinctions to sudden mutations and speciation events.

References

Almassy, N., Edelman, G. M., and Sporns, O. (1998). Behavioral constraints in the development of neuronal properties: a cortical model embedded in a real-world device. *Cerebral Cortex*, **88**, 1–16.

Bailey, C. H. and Kandel, E. R. (1993). Structural changes accompanying memory storage. *Annual Review of Physiology*, **55**, 397–426.

Bernstein, N. (1967). *The coordination and regulation of movements.* Pergamon, Oxford.

Blue, M. E. and Parnavelas, J. G. (1983). The formation and maturation of synapses in the visual cortex of the rat. II. Quantitative analysis. *Journal of Neurocytology*, **12**, 697–712.

Bourgeois, J.-P. and Rakic, P. (1993). Changes of synaptic density in the primary visual cortex of the macaque monkey from fetal to adult stage. *Journal of Neuroscience*, **13**, 2801–20.

Bourgeois, J.-P., Goldman-Rakic, P. S., and Rakic, P. (1994). Synaptogenesis in the prefrontal cortex of Rhesus Monkeys. *Cerebral Cortex*, **4**, 78–96.

Bradley, N. S. and Bekoff, A. (1990). Development of coordinated movement in chicks: I. Temporal analysis of hindlimb muscle synergies at embryonic days 9 and 10. *Developmental Psychobiology*, **23**, 763–82.

Calford, M. B. and Tweedale, R. (1990). Interhemispheric transfer of plasticity in the cerebral cortex. *Science*, **249**, 805–7.

Callaway, E. M. and Katz, L. C. (1991). Effects of binocular deprivation on the development of clustered horizontal connections in cat striate cortex. *Proceedings of the National Academy of Sciences, USA*, **88**, 745–9.

Cazalets, J. R., Menard, I., Cremieux, J., and Clark, F. (1990). Variability as a characteristic of immature motor systems: an electromyographic study of swimming in the newborn rat. *Behavioural Brain Research*, **40**, 215–25.

Cotman, C. W. and Nieto-Sampedro, M. (1984). Cell biology of synaptic plasticity. *Science*, **225**, 1287–94.

Darwin, C. (1859/1962). *The origin of species.* Collier, New York.

Dinse, H. R., Recanzone, G. H., and Merzenich, M. M. (1993). Alterations in correlated activity parallel ICMS-induced representational plasticity. *NeuroReport*, **5**, 173–6.

Dobzhansky, T. (1967). On types, genotypes, and the genetic diversity in populations. In *Genetic diversity and human behavior* (ed. J. N. Spuhler), pp. 1–18. Aldine, Chicago.

Edelman, G. M. (1987). *Neural Darwinism.* Basic Books, New York.

Edelman, G. M. (1988). *Topobiology.* Basic Books, New York.

Edelman, G. M. (1989). *The remembered present.* Basic Books, New York.

Edelman, G. M. (1992). *Bright air, brilliant fire.* Basic Books, New York.

Edelman, G. M. (1993). Neural Darwinism: selection and reentrant signaling in higher brain function. *Neuron*, **10**, 1–20.

Edelman, G. M. (1995). Memory and the individual soul: against silly reductionism. In *Nature's imagination* (ed. J. Cornwell), pp. 200–6. Oxford University Press, Oxford.

Edelman, G. M., Reeke, G. N., Gall, W. E., Tononi, G., Williams, D., and Sporns, O. (1992). Synthetic neural modeling applied to a real-world artifact. *Proceedings of the National Academy of Sciences, USA*, **89**, 7267–71.

Elbert, T., Pantev, C., Wienbruch, C., Rockstroh, B., and Taub, E. (1995). Increased cortical representation of the fingers of the left hand in string players. *Science*, **270**, 305–7.

Farde, L., Hall, H., Pauli, S., and Halldin, C. (1995). Variability in D-2-dopamine receptor density and affinity—A PET study with (C-11) raclopride in man. *Synapse*, **20**, 200–8.

Fodor, J. A. (1983). *The modularity of mind*. MIT Press, Cambridge, Mass.

French, A. S., Klimaszewski, A. R., and Stockbridge, L. L. (1993). The morphology of the sensory neuron in the cockroach femoral tactile spine. *Journal of Neurophysiology*, **69**, 669–73.

Greenough, W. T. and Chang, F. F. (1988). Plasticity of synapse structure and pattern in the cerebral cortex. In *Cerebral cortex*, Vol. 7. *Development and maturation of cerebral cortex* (ed. A. Peters and E. G. Jones), pp. 391–440. Plenum Press, New York.

Heisenberg, M., Heusipp, M., and Wanke, C. (1995). Structural plasticity in the *Drosophila* brain. *Journal of Neuroscience*, **15**, 1951–60.

Huttenlocher, P. R. (1979). Synaptic density in human frontal cortex—developmental changes and effects of aging. *Brain Research*, **163**, 195–205.

Johnson-Laird, P. N. (1988). *The computer and the mind*. MIT Press, Cambridge, Mass.

Lashley, K. S. (1947). Structural variation in the nervous system in relation to behavior. *Psychological Review*, **54**, 325–34.

Loeb, G. E. (1993). The distal hindlimb musculature of the cat—interanimal variability of locomotor-activity and cutaneous reflexes. *Experimental Brain Research*, **96**, 125–40.

Löwel, S. and Singer, W. (1992). Selection of intrinsic horizontal connections in the visual cortex by correlated neuronal activity. *Science*, **255**, 209–12.

Mai, J. K., Berger, K., and Sofroniew, M. V. (1993). Morphometric evaluation of neurophysin-immunoreactivity in the human brain: pronounced interindividual variability and evidence for altered staining patterns in schizophrenia. *Journal für Hirnforschung*, **34**, 133–54.

Mayr, E. (1959). Darwin and the evolutionary theory in biology. In *Evolution and anthropology: a centennial appraisal* (ed. B. J. Meggers), pp. 1–10. The Anthropological Society of Washington, Washington, D.C.

Mayr, E. (1963). *Animal species and evolution*. Belknap Press, Cambridge, Mass.

Merzenich, M. M., Nelson, R. J., Stryker, M. P., Cynader, M., Schoppman, A., and Zook, J. M. (1984). somatosensory cortical map changes following digit amputation in adult monkeys. *Journal of Comparative Neurology*, **224**, 591–605.

Merzenich, M. M., Nelson, R. J., Kaas, J. H., Stryker, M. P., Jenkins, W. M., Zook, J. H., *et al.* (1987). Variability in hand surface representations in areas 3b and 1 in adult owl and squirrel monkeys. *Journal of Comparative Neurology*, **258**, 281–96.

Pearson, K. G. and Goodman, C. S. (1979). Correlation of variability in structure with variability in synaptic connections of an identified interneuron in locusts. *Journal of Comparative Neurology*, **184**, 141–65.

Pons, T., Preston, E., Garraghty, A. K., Kaas, J., and Mishkin, M. (1991). Massive cortical reorganization in primates. *Science*, **252**, 1857–60.

Putnam, H. (1988). *Representation and reality*. MIT Press, Cambridge, Mass.

Purves, D. (1988). *Body and brain*. Harvard University Press, Cambridge, Mass.

Pylyshyn, Z. W. (1984). *Computation and cognition: toward a foundation for cognitive science*. MIT Press, Cambridge, Mass.

Quartz, S. R. and Sejnowski, T. J. (1997). The neural basis of cognitive development: a constructivist manifesto. *Behavioral and Brain Sciences*, **20**, 537–96.

Rajkowska, G. and Goldman-Rakic, P. S. (1995). Cytoarchitectonic definition of prefrontal areas in the normal human cortex: II. Variability in locations of areas

9 and 46 an relationship to the Talairach coordinate system. *Cerebral Cortex*, **5**, 323–37.

Ramachandran, V. S. (1993). Behavioral and magnetoencephalic correlates of plasticity in the adult human brain. *Proceedings of the National Academy of Sciences, USA*, **90**, 10413–20.

Recanzone, G. H., Merzenich, M. M., and Dinse, H. R. (1992). Expansion of the cortical representation of a specific skin field in primary somatosensory cortex by intracortical microstimulation. *Cerebral Cortex*, **2**, 181–96.

Reeke, G. N., Jr and Sporns, O. (1990). Selectionist models of perceptual and motor systems and implications for functionalist theories of brain function. *Physica D*, **42**, 347–64.

Reeke, G. N., Jr and Sporns, O. (1993). Behaviorally based modeling and computational approaches to neuroscience. *Annual Review of Neuroscience*, **16**, 597–623.

Reeke, G. N., Jr, Finkel, L. H., Sporns, O., and Edelman, G. M. (1990). Synthetic neural modelling: a multi-level approach to brain complexity. In *Signal and sense: local and global order in perceptual maps* (ed. G. M. Edelman, W. E. Gall, and W. M. Cowan), pp. 607–707. Wiley, New York.

Searle, J. R. (1992). *The rediscovery of the mind*. MIT Press, Cambridge, Mass.

Singer, W. and Gray, C. M. (1995). Visual feature integration and the temporal correlation hypothesis. *Annual Review of Neuroscience*, **18**, 555–86.

Spengler, F. and Dinse, H. R. (1992). ICMS induced emergence of new skin field representations in rat somatosensory cortex. *Society of Neuroscience Abstracts*, **22**, 345.

Sporns, O. (1997). Deconstructing neural constructivism. *Behavioral and Brain Sciences*, **20**, 576–77.

Sporns, O. and Edelman, G. M. (1993). Solving Bernstein's problem: a proposal for the development of coordinated movement by selection. *Child Development*, **64**, 960–81.

Sporns, O. and Jenkinson, S. (1997). Potassium ion- and nitric oxide-induced exocytosis from populations of hippocampal synapses during synaptic maturation *in vitro*. *Neuroscience*, **80**, 1057–73.

Sporns, O. and Tononi, G. (1994). *Selectionism and the brain*. Academic Press, San Diego.

Thelen, E. and Corbetta, D. (1994). Exploration and selection in the early acquisition of skill. In *Selectionism and the brain* (ed. O. Sporns and G. Tononi), pp. 75–102. Academic Press, San Diego.

Thelen, E. and Smith, L. B. (1994). *A dynamic systems approach to the development of cognition and action*. MIT Press, Cambridge, Mass.

Tramo, M. J., Loftus, W. C., Thomas, C. E., Green, R. L., Mott, L. A., and Gazzaniga, M. S. (1995). Surface area of human cerebral cortex and its gross morphological subdivisions: *in vivo* measurements in monozygotic twins suggest differential hemisphere effects of genetic factors. *Journal of Cognitive Neuroscience*, **7**, 292–301.

Watson, J. D. G., Myers, R., Frackowiak, R. S. J., Hajnal, J. V., Woods, R. P., Mazziotta, J. C., *et al.* (1993). Area V5 of the human brain: evidence from a combined study using positron emission tomography and magnetic resonance imaging. *Cerebral Cortex*, **3**, 79–94.

Weinberg, S. (1995). Reductionism redux. *The New York Review of Books* October 5, 39–42.

Wolff, J. R., Laskawi, R., Spatz, W. B., and Missler, M. (1995). Structural dynamics of synapses and synaptic components. *Behavioral Brain Research*, **66**, 13–20.

....

CHAPTER 4

Consciousness—the interface between affect and cognition

BERNARD BALLEINE and ANTHONY DICKINSON

Contemporary theories of consciousness emphasize, almost exclusively, the cognitive aspects of mentality at the expense of the conative. And one can see why—it is in the service of a unified ontology. Contemporary cognitive science has rendered the material basis of cognition unproblematic: with the development of artificial symbol-manipulating devices that can demonstrably represent and reason, there is no longer a principled barrier to understanding natural cognition in terms of brain processes. In order to ground the whole of the mental in the material, therefore, all that has to be assumed is that consciousness is but a species of cognition. But if there is general agreement that cognitive processes can be reduced to brain processes and that consciousness is a cognitive process, why should there be any disagreement over the obvious conclusion from these premises?

What is consciousness for?

It has been increasingly recognized that the cognitive hegemony established by this line of reasoning is but a finesse. For example, Chalmers (1996) argued that, although many 'easy' problems associated with consciousness (such as discrimination, categorization, the integration of information,or the reportability of mental states) can, in principle, be solved by focusing on the relation between neural and cognitive states, this focus will not solve the hard problem—the problem of our 'experience' of mental states, or what it feels like to be conscious. So the difficulty in accepting reductive approaches to consciousness appears to be that the premises of our syllogism are incomplete; the nature of consciousness is not to be found solely in the cognitive domain. But if we are to take seriously the view that explanations of consciousness must give an account of our 'experience' of mental states or of our 'experience' of what it is like to be conscious, then surely we must address how and why it is that, in addition to its other roles, consciousness functions as an organ of feeling.

An approach to this problem emerges in discerning how the hard and the easy problems differ. For Chalmers the easy problems are easy precisely because they concern the explanation of cognitive abilities and functions. And so, by implication, the hard problem could become an easier one if a function could be found for 'experience': namely for why we have evolved the capacity to experience our mental states just as we can enjoy the feeling of being conscious. Then, at least potentially, the problem of conscious experience could become one with which cognitive science could grapple.

The evolution of consciousness

Asking what something is for and why it evolved are related questions but not identical (cf. Brandon 1990). Conscious inessentialists (see, for example, Flanagan and Polger 1995) present the position that, although consciousness may have evolved it may have done so only because it was compresent with some other capacity on which evolution could act. Consciousness could, therefore, be a useless by-product of something that is useful. If that is so, then, as Davies and Humphreys (1993) have put it, so much for consciousness: 'psychological theory need not be concerned with this topic' (pp. 4–5). But if we deny the inessentialists' claim and shoulder the burden of providing a function for consciousness, then we are implicitly arguing that consciousness is an adaptation—something that gave those with it an advantage over their unconscious brethren and other competitors.

Several hypotheses about what consciousness is for and why it evolved have been entertained. This is not the place to describe them in detail but they fall, generally, into two groups: those that ascribe one or other basic cognitive function to consciousness (for example, Baars 1988; Crick and Koch 1990; Dennett 1991) and those that ascribe a more complex socio-cognitive or, related, linguistic function (e.g. Jerison 1973; Jaynes 1976; Humphrey 1983). Against the first, conscious inessentialists have directed their most telling blows. If a cognitive capacity can be utilized without consciousness, then its conscious deployment is unlikely to confer any additional advantage. Hence, no fundamental function of consciousness is likely to be discerned from further analysis. Nor is this problem overcome by proposing that consciousness allows some form of higher-order or meta-representational act. As Dretske (1995) argued, these views confuse 'object' with 'act' awareness; there is nothing in the object of cognition that, in principle, determines whether a cognitive act will be conscious or unconscious. Thus, even if the object of cognition is another act of cognition why should consciousness necessarily accompany this cognitive act but not others?

Against the socio-cognitivist approach to consciousness, the usual argument is that, if consciousness evolved as a means to engage in social activity, why is it that so much of our conscious life is taken up with very basic cognitive and sensory activities? Surely we should be more likely to become conscious of social and societal relations than simple perceptual features of the landscape? Yet this does not appear to be so (compare Humphrey's, 1992, arguments with his own earlier views).

Common to all of these approaches is a failure to broach Chalmers's hard question; there is no currently proposed function for conscious experience—indeed, some theorists explicitly deny that experience plays any functional role at all (for example, Dennett 1991). It may be that the current addiction to the computer analogy of brain function has made it difficult to consider qualitative issues, but, along with the bath water, a baby seems to have gone missing: a most curious omission from almost every current view of consciousness is the role that this capacity plays in the experiencing of affects or emotions.

The same cannot be said of dynamic depth psychology, of which psychoanalysis is the best known example. Freud, for example, was a strong advocate of the position that most of our mental life is unconscious and argued this position on the basis of the results of his clinical studies of psychopathology. Nevertheless, he clearly regarded consciousness as serving a distinct and important function and, where possible, was keen to point to the principle distinctions between conscious and unconscious mental acts. The clearest statement of a distinction of this kind can be found in his paper 'The unconscious' (1915). In the second part of that paper he assesses whether affects or emotions, central to his theory of repression, can be deemed to be unconscious processes in similar fashion to unconscious mental acts. Initially, he suggests: 'It is surely of the essence of an emotion that we should be aware of it, i.e. that it should become known to consciousness. Thus the possibility of attributing the processing of emotions, feelings and affects to the unconscious is completely excluded' (p. 179). After examining potential counterarguments derived from the vicissitudes in affective states observed to accompany certain neuroses, he nevertheless concludes: 'There are no unconscious affects as there are unconscious ideas' (p. 180).

In contrast to the current focus on cognitive processing, Freud's analysis of the clinical material suggests an alternative and important function to consciousness—that it is necessary for the presentation of affect. But, it will be countered, surely there is nothing novel in this suggestion. The function that conscious lusts and pains play in determining approach and avoidance behaviours has been thought to be central to adaptive behaviour before now. But where these approaches have fallen is in their failure to suggest what advantage the *conscious* expression of these emotional states confers over and above their unconscious biological correlates. It is to this critical question that we now turn.

Emotion and action

It is not usually a problem convincing people that they have two modes of acting—one relatively reflexive, automatic, and under little if any 'executive' control, and another deliberate, intended and voluntary. It is difficult, for example, to salivate on command although we can do so if we imagine biting into a lemon (but now try to stop salivating). Although we often cannot directly control them, conditioned reflexes or Pavlovian conditioned responses (of which salivation elicited by our hallucinatory lemon is but one example) offer an animal considerable advantages. Sensitivity to events and to signals for events that are of primary biological significance for survival can allow an animal to behave in a manner that anticipates contact with food or water or a mate as well as a predator. Such anticipatory behaviour confers advantages over other animals that must actually come into direct contact with events before these behavioural responses are elicited. But if these reflexive reactions to signals for events of significance offer an advantage, how much more of an advantage must have been conferred by the ability to *control* one's actions—to survey the scene and bring other sources of information to bear on the problem of deciding on a course of action without reacting in an involuntary, reflexive manner. Consider what advantage it must have conferred, in an environment with scarce resources, to be able to turn a predator into a meal rather than an opportunity for a little unplanned exercise.

There is, however, a serious problem that must have faced animals during the development of this more controlled mode of behaviour for, if the adaptive nature of one's reflexive behaviour is guaranteed by natural selection, there is no guarantee that one's intentional actions will be so adaptive. Once we can control our actions we are faced with a dilemma; as Dennett (1991) puts it: 'Now what do I do?' (p. 177). What will I pursue and what avoid? In short, the problem to be resolved is where do the goals of our actions come from? If intentional actions are to remain adaptive it is clear that they must have consequences that enhance survival. But, to pursue those kinds of consequences, it is necessary to monitor the current state of our biological system, a system of motivational needs that is opaque to our five senses. Only then could an intentional system learn to pursue those things that are needed and avoid things, including plants, that may damage or kill.

It is our thesis that, to achieve those ends, the capacity to externalize biologically based motivational states evolved. But to be cognisant of these states, and so integrate them with environmental contingencies to derive a course of action, required that these states be represented in a manner available to a cognitive system. This, we argue, was achieved by the development of a new capacity, the capacity to reflect the activity of internal biologically based motivational states as conscious feelings or emotions.

Conscious versus unconscious desires

Our view on consciousness, which will be progressively justified below, is that it evolved to reflect biological states as emotions so to allow these to be associated, in a cognitive system, with the representations of external events to form evaluative beliefs or *desires*. From this perspective, therefore, there are, strictly speaking, no unconscious emotions although, as evaluative beliefs, there can clearly be unconscious desires, a position that accords with the psychoanalytic view of psychopathology.

A problem that many theories of consciousness come up against is that the central importance of consciousness to adaptive behaviour is sometimes lost in the seemingly trivial function that it is deemed to serve. But it should be clear that there is nothing trivial about the role that the development of conscious emotions serves, if these reflect the current state of the biology. Science-fiction movies (and not a few philosophical manuscripts) are full of examples of creatures, some biological, some cybernetic, some both, who seem to be devoid of sensitivity to the consequences of their actions, with the predictable catastrophic results for a puny but determined humanity. For most writers this scenario serves as no more than a warning about the dangers of a lax morality or an unfettered faith in technology. But what of the creatures themselves? For whom could the development of intentional action be more dangerous than for a creature incapable of observing that the consequences of its actions threatened its own existence?

To develop our thesis and its implications, we begin, therefore, by considering various approaches to goal-directed action. This will allow us to clarify the kind of problem that the development of this mode of behaviour presented to evolution and the evidence that, we believe, suggests that this problem was solved by the development of conscious emotional states. We then consider the evidence for the relation between emotional and biological states, and the way in which the value of goals is conferred, before coming back to our thesis—to make plain the way we believe conscious emotions interact with a cognitive system to determine a course of action. Finally, we offer a novel prediction and report an experiment that tests this prediction.

Goal-directed action

The adaptive significance of the capacity for goal-directed action should warrant little comment; it is this capacity that allows us and other animals to control the environment in the service of our basic needs and desires. In addition, it should be immediately apparent that, at least from a functional perspective, the deployment of cognitive resources should be closely related to this capacity. If we were unable to modify our behaviour on the basis of knowledge that we gain of changes in environmental events, this knowledge could never be expressed and would remain bound within the mind, rendering us silent spectators of our transactions with the world. Imagine a biological system able to cognize environmental events and their relations but unable to control its actions in response to those events. Such a system would have to rely on evolution, rather than its cognitive apparatus, to equip it with the behavioural reactions suited to survival—those that allowed it to gain essential nutrients and avoid sources of danger. And even if nature equipped this system with sufficient flexibility to react to signals that predict biologically important events in addition to the events themselves, still this system would be reliant on environmental stability if these reactions were to remain adaptive. (We have broached the problem of how actions differ from non-intentional, behavioural reactions and how they may be differentiated empirically elsewhere and do not consider this issue further here; cf. Dickinson and Balleine 1993, 1994.)

It is no accident, therefore, that identifiably cognitive explanations of our actions have been with us for a long time. In the late nineteenth century the arguments of the empiricist–associationist tradition—that the mind is essentially passive—strongly influenced the then new psychology, which hence focused experimental attention on the contents, as opposed to the act, of cognition. However, experimental evidence forced various investigators, particularly those belonging to the Wurzburg School, to acknowledge that thought was active, that it had direction and orientation. These theorists drew a strong relation between the activity of thought and that of overt behaviour by proposing that both were determined by their orientation towards an objective or goal. This relation formed the basis of purposive theories of action by proposing that actions were determined by the purpose or goal that the agent had in mind.

In the early twentieth century gradual recognition of the failure of cognitive approaches and of the introspective methodology to supply consistent and useful observations led to the introduction of definitions of purpose linked to observable behaviour rather than to mental processes. The earliest such programme was developed by William McDougall (1912, 1925), who, in distinguishing purposive action from other behaviour, such as reflexes, proposed a list of the 'marks of purpose', beginning with the fact that purposive behaviour displayed 'a certain spontaneity of movement' and then elaborating this characterization with each succeeding mark: thus, the agent persists until it gets to the goal; varies its direction to avoid obstacles; ceases its movements when, and not until, they result in the attainment of the goal; moves to anticipate or prepare for changes in the situation to come; shows similar but improved movements when a situation is again encountered; and acts as a totality in achieving its end.

In defining purposive behaviour in this way McDougall may have been seeking to formulate an objective approach, but it is clear that this was not a programme he was prepared to apply consistently. Indeed, in line with his view that 'we may define the mind of any organism as the sum of the enduring conditions of its purposive activities' (McDougall 1912, pp. 69–70), he saw in the spontaneous, persistently goal-oriented nature of the actions of an agent evidence for purpose as a distinct determinant. Purpose became for McDougall something distinct from behaviour, what he referred to as a 'hormic energy'. This attempt to define purposes as an internal force in this way was really no more than a sleight of hand by which McDougall hoped to avoid the problem of explaining why his purposes, that were really no more than a description of behaviour–environment contingencies, could induce the agent to produce an overt action. In effect, it offered no advance over previous cognitive approaches because the agent was, in practice, left to initiate its own behaviour, direct itself around obstacles and choose between actions solely on the basis of its beliefs concerning the consequences of those actions.

Nevertheless, McDougall's externalist approach can be found embedded in several current psychological (see Frese and Sabini 1985) and philosophical (Taylor 1970) theories of action. Whatever the details of these more recent formulations, however, similar problems have arisen associated with the emphasis placed on the rationality of the performance of an action with respect to its consequences; that, in some way, beliefs about the consequences of an action can alone determine a course of action. Borger (9170) offers a clear critique of this position, suggesting that, in a given situation, explaining why a particular action is performed at a particular time is not possible with reference to beliefs about the consequences alone; that informational states cannot in and of themselves determine a course of action. In essence, Borger's contention is that any information that takes the form 'action A leads to consequence C' does not imply a course of action because this information may be used both to perform A or to avoid performing A. For example, if C is food, this belief should lead us to perform A when hungry but not when satiated. By contrast, anorexics may avoid performing A whatever their nutritional demands. In deciding why and when an agent will prefer one action over another we are forced, therefore, to introduce some basis for choosing between different actions, for otherwise our answer to the question 'why did the agent perform action A rather than action B?' can be no more than 'because they chose to'. Unfortunately, the introduction of volitional states, such as choice or free will, by implying that behaviour is intrinsically unpredictable, offers no basis for a scientific approach to goal-directed action.

The problem of desire

Alternatively, it is possible to formulate accounts of behaviour that, in addition to beliefs about the results of an action, suppose that choice between actions is determined by internal properties of the agent sensitive to the desirability of the consequences. Thus, in order to account for why the agent performed action A, these theories contend that the agent must not merely believe that 'A leads to C' but they must also 'desire C'. Woodfield (1976) offers an example of this account of action. Our answer to the question why does the agent do B rather than A is given in these terms: 'S did B because S wanted to do G and believed that B promoted G' (p. 182). From this perspective the instrumental belief: 'S believes that B promotes G' is coupled with a desire 'S wants to do G' to derive, through practical inference, an intention to do B.

Although the causal status of intentions has been questioned in the past (Maze 1983; cf. Mackie 1974), the current attitude is aptly summarized by Pettit (1979) when he states: 'The question of status appears to be settled. The explanation of actions by reference to states of belief and concern [i.e. intention or desire] represents those actions as the determined causal issue of the mental states' (p. 8). It is not, then, an exaggeration to state that the intentional approach is now the dominant theoretical approach to action. Nevertheless, although it is now acceptable to contend that agents are sensitive to the consequences of their actions and, indeed, to formulate extensive research programmes to explore this aspect of action, little if any detailed consideration has been given to the nature of desires. Given that this aspect of current theory must ultimately differentiate it from less-testable alternatives, we may feel justified in asking: 'What is the nature of desire and what is its source?'

Within the intentional perspective, it is usual to suggest that the existence of goals is sufficient to define what is meant by desire. Again, Woodfield (1976) makes this point explicit: 'The most obvious kind of mental state with which to identify the state of having a goal is the state of wanting or desiring ... The core-concept of a goal is the concept of an intentional object of desire' (p. 166).

As this statement attempts to make clear, cognitive approaches to action suppose that the desire for a particular goal is itself a cognitive act—desiring always involves desiring a *thing* of a particular kind. Although this approach gives a unified cognitive account of action it has, nevertheless, left a rather awkward gap in theory, suggesting, as it does, that all the goals of our actions are acquired through cognitive acts. This would appear to be an un-exceptionable suggestion in the case of social goals such as truth, justice, or the American way, the desirability of which must certainly be a matter of learning, but how do we account for actions directed towards attaining goals, such as foods and fluids, that have biological utility? Do we need to learn that commodities with biological utility are desirable through basic cognitive processes before they can act as the goals of action? Surely this is an untenable supposition of the theory.

This discussion has identified a central issue in any thorough-going account of action: that is, the role that biologically active motivational processes play in our psychological life. It is difficult to see how evolutionary pressures could have forced the development of cognition and action solely for the commerce of social goods. Nevertheless, the role these functions play in the immediate control of physiological systems must surely be limited. It is clear that we have only limited direct knowledge of the state of our physiological systems. It is also clear, however, that the goals of our actions must in some way be based in the needs of our biological systems for we would soon cease to function at all without vital supplies.

The problem that evolution had to solve in order that intentional action could develop was, therefore, to establish the means by which a cognate system could ground the goals of intentional actions in physiological states. How we believe this was achieved will shortly occupy us in some detail. Explicating the processes that determine the performance of actions that obtain commodities of biological utility is, however, an important step in our argument and so we shall first turn to that issue to establish the basis for our subsequent claims.

Motivational control of goal-directed action

Although it is possible to observe the effect of changes in biological conditions on human action it is rare that such studies are conducted because, in these situations, attempts to undertake the carefully controlled study of human subjects is extremely difficult. Humans are rarely willing to put up with the experimental manipulation of primary goals for the sake of psychological investigation. As a consequence, the study of the motivational control of goal-directed action has, in the main, been conducted using animals other than humans as subjects. The evidence we shall present in this section has come mainly from our laboratories where we have used rats as subjects.

It is well established that, in rats, the biological conditions induced by the deprivation of specific commodities—what we call specific motivational states—are a major determinant of their goal-directed action; not surprisingly, hungry animals perform actions that gain access to a nutrient outcome more vigorously than sated ones. But current evidence suggests that this is not because deprivational states directly modulate the goal value—what we call the incentive value—of biologically relevant commodities. Rather, as suggested by the cognitive approach to action presented above, it appears that animals have to learn about the desirability of commodities when in a particular motivational state before they will modify the performance of actions to gain access to those commodities. This evidence suggests, therefore, that animals are more likely to pursue a course of action that leads to food when hungry than when sated because food-deprivation induces animals to assign a higher incentive value to nutritive outcomes when they are consumed in the hungry state (for a review, see Balleine 1992; Dickinson and Balleine, 1994).

Evidence for this suggestion comes from studies assessing the effects of shifts in motivational state on animal action. For example, Balleine (1992) trained groups of non-deprived rats to press a freely available lever to get food pellets that the animals had never before experienced. After training, half of the rats were shifted to a food-deprived state, group DEP, while the remainder were maintained non-deprived, group NON, before both groups were given a test on the levers conducted in 'extinction'; that is, in the absence of the food pellets. It is important to note the reason for using an extinction test. If Balleine had delivered the pellets during the test any difference between deprived and non-deprived animals may have been due to learning that occurred on test rather than the effects of the shift in motivational state. If the biological conditions induced by depriving the rats of food immediately and directly increases the desirability of nutritive commodities, then we should expect food-deprived rats to press the lever more than non-deprived rats in this test. As is clear from Fig. 4.1a, however, Balleine found that performance of the groups on test did not differ. Nevertheless, the shift in motivational state was clearly effective. As can be seen from Fig. 4.1b, in a subsequent test where the animals could again earn food pellets by lever pressing, the food-deprived rats pressed at a substantially higher rate than the non-deprived rats. Hence, lever-press performance was only affected by the shift in motivational state if the rats were allowed contact with the pellets after a lever press.

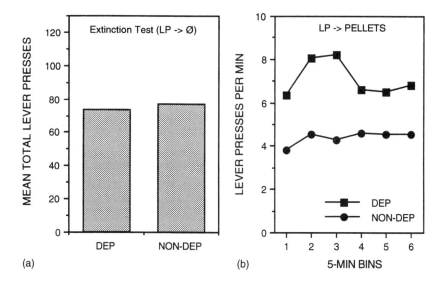

Fig. 4.1 The results of an experiment assessing the effects of a post-training increase in food deprivation on lever-press performance. Animals were trained to lever press for the pellets when non-deprived. One group was then deprived of food (DEP) whereas the remainder were maintained non-deprived (NON-DEP). Animals were then given two tests on the levers; the first (a) was conducted in extinction (i.e. LP → Ø), whereas, in the second (b), lever pressing again earned the food pellets (LP → pellets).

Careful consideration of this pattern of results suggests that an important determinant of the effect of a shift in motivational state on the performance of actions is consummatory contact with the outcome in the new motivational state. Perhaps because the rats had never before experienced the food pellets when food deprived they were unaware of their increased value when in that state. In subsequent experiments (for example, Balleine 1992; Balleine *et al.* 1994) we have directly assessed this prediction using four groups of rats, all trained to lever press for food pellets when undeprived. After training, half of the rats were to be tested in a food-deprived state whereas the remainder were to be tested in the non-deprived state. Before the test, however, half of each of these test groups were allowed to consume the food pellets in separate feeding cages whilst in a food-deprived state. The remainder were allowed to eat the food pellets in the non-deprived state. This generated four groups: one tested food deprived and food deprived after re-exposure to the pellets (DEP-DEP); one tested food deprived but exposed to the pellets only when non-deprived (DEP-NON); one tested non-deprived but previously exposed when food deprived (NON-DEP); and, finally, a group both tested and exposed when non-deprived (NON-NON). If consummatory experience with a novel food in a food-deprived state is necessary for that state to influence the performance of actions that gain access to that commodity, then animals in group DEP-DEP who are given that experience should increase their lever pressing in the extinction test relative to the other three groups. As is clear from Fig. 4.2, this is indeed just the pattern of results that Balleine *et al.* (1994) observed.

In further experiments, we were able to confirm that this effect of consummatory contact with the outcome depended on the particular action–outcome relation that was revalued. In these experiments non-deprived rats were trained to perform two actions, lever pressing and chain pulling; one action earned access to food pellets and the other to a solution of maltodextrin, a complex starch. All rats were then given a choice extinction test on the levers and chains. Before the test, however, the animals were given sessions in which they were allowed to eat one outcome of the instrumental action when food

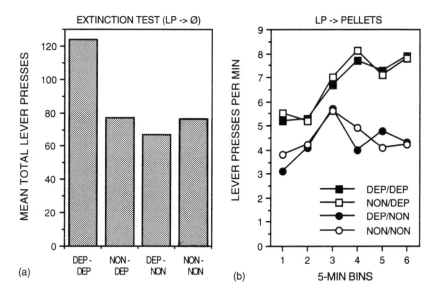

Fig. 4.2 The results of an experiment assessing the effects of incentive learning on performance after a post-training increase in primary motivation. Animals were trained undeprived to lever press for pellets. Two groups (DEP-DEP and NON-DEP) were then deprived of food whereas the remaining groups (DEP-NON and NON-NON) were maintained undeprived before lever pressing was assessed in extinction (a) and when pressing again earned the pellets (b). Before the test, animals in groups DEP-DEP and DEP-NON were given the opportunity for consummatory contact with the food pellets when food deprived whereas animals in groups NON-DEP and NON-NON were not given this opportunity.

deprived and, on alternate days, the other outcome in the training, non-deprived state. The results of the test phase are presented in Fig. 4.3

In accord with the suggestion that eating the outcome is needed to observe the effect of a shift in motivational state on the performance of actions, we found that animals performed more of the action that, in training, had delivered the outcome subsequent eaten in the food-deprived state before the test (DEP) than the other action (NON-DEP). But, more importantly, this study clearly and powerfully demonstrates that the actions of animals are determined by the integration of two sources of information: (1) beliefs concerning the consequences of actions acquired during training and (2) the current desirability of the outcome of these actions. If the rats were insensitive to any of these factors they could not have formulated a course of action in the extinction test. That rats are capable of doing so is strong evidence that they encode and integrate these sources of information. Furthermore, this study

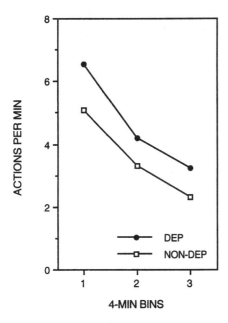

Fig. 4.3 The results of an experiment assessing the effects of incentive learning on the choice between two actions after a post-training increase in primary motivation. Animals were trained to lever press and chainpull with one action earning food pellets and the other a maltodextrin solution before being given a choice between the levers and chains when food deprived in an extinction test. Before the test, animals were allowed to consume one of the outcomes when food deprived and the other when non-deprived. Performance on test is presented separately for the action whose training outcome was contacted when food deprived (DEP) and for the action whose training outcome was contacted when non-deprived (NON-DEP) before the test.

demonstrates that the desirability of the outcome of an action is clearly not something directly determined by the motivational state alone. Rather, a desire is a relation in which a biologically defined motivational system stands with a particular event or commodity that is established through consummatory contact with that commodity.

It is worth pointing out the generality of the effects of consummatory contact with biologically relevant commodities on the performance of actions after a shift in motivational state, for this effect is not confined to post-training increases in food deprivation. The same pattern of results has also been found for the opposite shift—that is, where rats were trained to lever press for food pellets when food deprived and then tested when undeprived. In this case, rats reduced their performance when food deprivation was reduced only if they were allowed to eat the food outcome when undeprived before the test (Balleine 1992; Balleine and Dickinson 1994). Further, the generality of this effect has been confirmed for several different motivational systems and for several commodities relevant to those states. For example, similar effects have been found to mediate the effects of shifts in water deprivation (Lopez *et al.* 1992); changes in outcome value mediated by drug states (Balleine and Dickinson 1994; Balleine *et al.* 1994; Balleine *et al.* 1995*a*) changes in the value of goals relevant to thermoregulatory needs (Henderson and Graham 1979) and sexual needs (Everitt and Stacey 1987—see Dickinson and Balleine 1994 for a review). In all of these cases it is clear that animals have to learn about changes in the desirability of biologically relevant commodities through consummatory contact with them before a change in motivational state will affect the performance of actions. We call this process of learning about changes in the desirability of commodities through consummatory processing *incentive learning* (Dickinson and Balleine 1994).

Determination of incentive value

Although it is clear that motivational systems determine the incentive values assigned to particular commodities, it remains to be established how this is accomplished. One suggestion is that the incentive value is a product of establishing the commodity as a signal for a particular consequence. There is, indeed, evidence that animals can form preferences for commodities on the basis of the consequences that they predict. For example, presenting a hungry animal with food mixed with a particular taste will increase the preference for that taste even when it is subsequently presented without the previously accompanying nutrients. This, it is argued, is due to knowledge of the predictive relation between the taste and the post-ingestive consequences produced by eating food when hungry. This same effect can be established if, instead of mixing the flavour with the food, the flavour is first presented followed by the food after a delay, or even if the food is not ingested by the animal but is intubated directly into the stomach (Capaldi 1992).

In addition to these increments in the preference for a commodity paired with the repletion of food deprivation, aversions to particular foods and fluids can be induced by the gastric events associated with disgust. If an animal becomes sick after ingesting food, an aversion will be produced for that food—an effect referred to as taste-aversion learning (Garcia *et al.* 1966). In these demonstrations, rats are allowed to eat a palatable commodity, which is often, although not exclusively, a sweet-tasting solution. Shortly after the session in which they were exposed to the sweet taste, some of the rats are injected with an agent that induces illness, generally lithium chloride (LiCl). Even though it was the injection and not the sweet solution that causes illness, when the rats are again allowed to eat the sweet solution, those made ill after the first exposure will no longer eat it. It is sufficient, therefore that particular post-ingestive consequences are associated with ingestion of a commodity; these consequences do not have to be caused by the commodity itself.

Results such as these are often cited as evidence that the incentive value of foods and fluids may be modified by establishing them as signals for particular, generally post-ingestive, consequences. An alternative view, however, is that associating a commodity with a particular consequence changes the affective or hedonic reactions to the commodity itself. Evidence for this latter position may be drawn from a study by Baeyens *et al.* (1990), who gave human volunteers the opportunity to drink some distinctively coloured or distinctly flavoured solutions. In addition, on some occasions particular solutions contained an additive that made it taste particularly unpleasant. Thus, for one group of subjects the distasteful solution had a distinctive flavour, whereas for the other group it had a distinctive colour. After this training subjects were asked to give a rating of the attractiveness of the colour and the flavour cues themselves and then to state whether they had been previously paired with the distasteful solution. When the flavour cues were presented alone on test, there was a reduction in the ratings of the attractiveness of those flavours when they had been presented in the distasteful solution. Nevertheless, the subjects were unable to report reliably which flavours had been presented with the distasteful solution. This was not simply due to the insensitivity of this measure of the subject's knowledge of the flavour–distaste associations, for they were perfectly able to report which colours had been paired with the distasteful solution. Importantly, however, these pairings did not alter their ratings of the attractiveness of the colours themselves.

This study demonstrates that the value of a commodity is not solely mediated by knowledge of the consequences of consummatory contact but also by the hedonic or affective reactions elicited by such contact. The fact that signalling and affective processes were double dissociated by the type of cue in the Baeyens *et al.* (1990) study indicates that these processes are mediated by different learning mechanisms. Within the present context, the issue is whether, during an incentive learning experience, agents assign incentive value on the basis of their affective response to the outcome itself or on the basis of learning about the predictive or signalling relation between the consumption and its consequences.

Motivation and affect

Hedonic theories explain an agent's actions in terms of the role these actions play in maximizing pleasure and minimizing pain. This account argues, therefore, that incentive learning is determined by the hedonic or affective responses elicited by experience with the outcome. As a consequence, evaluation of this theory requires us to relate affective responses to the performance of actions across variations in motivational parameters, such as deprivational state. The problem with this strategy, however, is that the hedonic experience itself—that is, the positive or negative affective response—is revealed only indirectly and then in humans only by subjective reports such as those used by Baeyens and his colleagues. Indeed, the need for measures such as rating scales to evaluate hedonic experience points to a major problem in the evaluation of hedonic theories. In situations where no measure of hedonic experience, however indirect, is available, it is possible to gauge affective responses only on the basis of overt behaviour. Thus, for example, we might assume that a particular commodity is pleasant for a very young child or an animal if they approach and eat it or learn to perform more-complex responses to gain access to it, even though we have no independent means of assessing whether it is in fact pleasure that is responsible for these responses. The lack of an independent measure can, in addition, render explanations in terms of hedonic experience circular. For example, we may ask 'why did the rat press the lever for food?', to which the answer might be given 'because it finds the food pleasant'. But then we may ask 'how do you know that it finds the food pleasant?', to which the only answer available is 'because it pressed the lever'.

A possible escape from circularity is provided by the immediate reactions elicited by contact with foods or fluids. For example, in human infants, the ingestion of a sugar solution produces a marked change in facial responses, such as smiling and lip licking, whereas the ingestion of a bitter-tasting quinine solution produces different responses, such as mouth gapes and tongue protrusions accompanied by arm flails and even screams and wails. Accompanying the positive facial responses to sugar are immediate changes in the pattern of sucking and an increase in heart rate indicative of autonomic arousal. Of course, adult humans also reflect different emotions in facial expressions, many of which resemble the basic acceptance and rejection responses associated with consummatory responses.

Several authors have also reported that similar orofacial patterns are observed in rats (Grill and Norgren 1978; Grill and Berridge 1985). When rats, like humans, are given sweet-tasting solutions to drink, the responses they make are termed ingestive consummatory responses, and include paw licking, rhythmic mouth movements, and tongue protrusions. In contrast, when rats are given a bitter-tasting solution the responses they make are termed rejection responses, and include head shakes, chin rubs, and face washing. If these fixed-action patterns reflect the animal's hedonic response to a commodity, as these authors suggest, then they provide a vehicle for assessing the affective account of incentive learning. Variables that have an effect on incentive value, as assessed by the performance of particular actions, should have a corresponding effect on these affective or 'hedonic' responses.

As we have seen, primary among these variables is the agent's motivational state—hungry animals assign a higher incentive value to a food when hungry than when sated. Consequently, a hedonic account of incentive learning requires that the motivational state should modulate affective responses in a corresponding manner. Evidence that affect is modified by changes in motivational state may be derived from Cabanac's studies using human subjects (see Cabanac 1971). Cabanac asked two groups of subjects to rate how much they liked the taste of a sweet-tasting glucose solution. Both groups were allowed to taste 50 ml of solution ever 3 minutes. One group was asked to swallow the solution, the other was asked to spit it out. Initially, both groups gave high ratings of pleasure. By the time 1000 ml of solution had been consumed, however, the ratings of subjects asked to swallow the solution changed from 'highly pleasant' to 'highly unpleasant'. In the group that was asked to spit the solution out, the rating remained at 'highly pleasant' throughout the test. Thus, the subjects' positive affect reactions decreased not because of taste-related processes alone but as they became increasingly sated on the sugar solution.

The corresponding motivational modulation of affective responses can also be observed in rats. For example, the incidence of ingestive fixed-action patterns elicited by a sweet solution increases with the level of food deprivation (Berridge 1991). Moreover, if rats are given a prior intragastric infusion of 5 ml of glucose, their orofacial responses change from ingestive to those associated with rejection, just as the hedonic ratings of human subjects changed from positive to negative (Cabanac 1990). This pattern of results in rats can also be observed using deprivation states other than hunger. A hypertonic sodium chloride solution, more salty than sea water, produces mainly rejection responses in rats. If, however, the rats are deprived of salt to induce a sodium appetite the hypertonic solution now elicits an increase in ingestive responses and a reduction in rejection responses (Berridge *et al.* 1984). Sodium-depleted humans also report that salty foods are more pleasant (cf. Rolls 1990).

Finally, an affective account can also been given of the loss of incentive value that follows the conditioning of an aversion to a flavour. Interestingly, when rats are made ill after eating a sweet-tasting solution, the responses elicited by a sweet taste change from being those associated with ingestion to those associated with rejection, suggesting that the affective valence of the solution itself has changed (Berridge *et al.* 1981). Traditionally, taste-aversion learning has been viewed as an example of Pavlovian conditioning with reduced consumption of the taste stimulus being thought to be induced by the formation of an association between the taste and illness. More recently, Garcia (1989) has proposed that taste aversions are mediated not only by the taste–illness association but by a change in the incentive properties, or palatability, of the taste. Thus, in line with findings that taste-reactivity responses to a solution change from being ingestive to being, after poisoning, indicative of rejection, Garcia's claim is that animals reject tastes paired with illness when they are subsequently contacted because they taste noxious. From this perspective the taste–illness association is latent in the physiology and only becomes manifest when the animal is exposed to the poisoned taste, during which exposure it discovers that the incentive value of the outcome has changed.

Arguments such as these have led some researchers to propose that the hedonic or affective experience elicited by a particular commodity reflects its 'usefulness' to current internal needs. Affect, from this perspective, is linked to physiological state in such a way that particular commodities produce pleasure when they reduce a particular 'need'. Indeed, the affective response to commodities that satisfy internal needs appears to be so adaptive that researchers have long suggested that this relationship between affective and physiological states reflects a 'wisdom of the body' (Cannon 1947). Thus it is suggested, our desire for specific commodities does not reflect the vagaries of our whims and wishes but the actual state of our internal physiological systems.

Affect and incentive learning

Given that incentive value can be determined by an agent's affective reaction to an outcome and that motivation states act to modulate these reactions, it is clear why an incentive learning experience is necessary for a motivational shift to alter goal-directed action. In the absence of any experience of their affective reactions to the outcome in a particular motivational state, agents have no basis on which to assign a new and appropriate incentive value. Thus, according to this affective account, incentive learning consists of learning about one's affective reaction to a commodity when in a particular motivational state and making an assignment of incentive value on that basis.

The affective account of incentive learning has been tested using taste-aversion procedures to modify the value of the outcome rather than a shift in motivational state. For example, Balleine and Dickinson (1991) trained thirsty rats to lever press and chain pull in a single session with one action earning access to a sucrose solution and the other to a salt solution. Immediately after this session animals were injected with LiCl in the hope of conditioning a taste aversion to both of these outcomes. Next day, the animals were allowed merely to drink a small quantity of one of the two outcomes, the aim being that this treatment would allow them to discover that the incentive value of that outcome had changed, before they were given a choice extinction test on the levers and chains. If, in taste-aversion learning, animals learn merely about the predictive or signalling relation between a taste and illness, re-exposure to an outcome after the illness should, if anything, partially extinguish that association and hence result in *increased* performance of the action that earned that outcome in training. If, however, Garcia's (1989) analysis is correct, the incentive learning treatment should have allowed the animals to learn about the reduced incentive value of the re-exposed outcome relative to the other outcome, which, when integrated with the action–outcome relations encoded during training, should then be manifest on test in *reduced* performance of the action that earned the re-exposed outcome. This latter prediction is exactly what Balleine and Dickinson (1991) found. As predicted by Garcia's (1989) approach, in the choice test animals performed fewer of the action that, in training, delivered the outcome to which they were re-exposed before the test.

The significance of this finding is twofold. First, it confirms that, in similar way to other motivational systems, incentive learning plays a role in the way disgust, as a defensive motivational system, controls goal-directed action. Second, this study also allows us to conclude that incentive learning is not a form of signal learning but a means by which animals learn about a change in their affective reactions to a commodity. Because these reactions are elicited only when consummatory responses are performed on commodities, changes in incentive value that accompany a change in motivational state must be discovered through experience. Only through experiencing a change in incentive value is an animal's desire for a particular outcome or goal modified.

The process of incentive learning allows animals to base their desires on their affective reactions. This position suggests, therefore, that desires are not immediately determined or controlled by motivational systems but are a product of the relation between those systems and particular commodities established when animals engage in consummatory contact with those com-modities. Thus, although the value of goals with biological utility must be learned, what this analysis suggests is that desiring is not a simple cognitive act. Rather, desiring is derived indirectly from the experience of an affective response elicited when a particular need state comes into relation with quite specific environmental events.

Affect, cognition, and the formation of goals

The above discussion was initiated by what we referred to as the problem of desire; that is, how does a cognisant animal base the goals of its actions on biological needs, and so pursue goals that have biological utility, if it is unable directly to perceive those needs. The evidence we have discussed clearly indicates that the desirability of goals with biological utility must be learned. But, also, it is clear that the desirability of these goals is not based on a direct form of cognitive evaluation. Rather, discovering the desirability of particular commodities is based on the effects of consummatory contact with that commodity. This, we argued, is because commodities with biological utility—those that reduce or modify need states or visceral reactions—are productive of a particular and important effect in the animal; an affective response. This response allows the formulation of desires appropriate to biological needs without the agent being required to perceive those needs directly.

We contend, therefore, that the need to ground the goals of cognitively mediated actions in biologically relevant processes was solved by the evolution of an interface between these biological processes and cognition, namely con-sciousness. This interface enables physiological states, through their capacity to modulate affective reactions, to determine the desirability of specific objects and events and hence the acquisition of beliefs about their value as a goal. These evaluative beliefs are abstract in the sense that, once assigned, they are cognitively realized and, in the absence of further opportunity for incentive learning, will not be influenced by changes in the biological conditions that originally determined them. The unified cognitive nature of this account allows beliefs concerning the consequences of an action and beliefs concerning the value of those consequences to be integrated to derive a course of action.

Of course there is much that remains to be spelled out in this account. But, at least at a behavioural level, this position is sufficiently well articulated to provide the basis for prediction and empirical evaluation. Indeed, we have begun to test some of the implications that arise from this view. One direct prediction is that, once animals have experienced their affective response through consummatory contact with a commodity, and so have, presumably, assigned an abstract goal value to that commodity, changes in the physiological state that determines the assigned value should not influence subsequent performance in the absence of any further opportunity to learn about the effects of those changes. In one test of this prediction, we utilized a variant of the procedure described in the previous section. A conditioned taste-aversion procedure was used to devalue the outcome of an action (the design of this experiment is presented in the top panel of Table 4.1). We (Balleine *et al.* 1995*b*) trained thirsty rats to lever press and chain pull with one action earning access to sucrose solution and the other earning access to a salt solution. Immediately after this training we injected the animals with LiCl in the hope of conditioning an aversion to both solutions. In the next phase of the experiment we gave animals the opportunity for consummatory contact with both of the outcomes. Before exposure to one of the outcomes the animals were injected with physiological saline that should not have influenced their evaluation of the outcome and so, from our previous studies, we predicted that contact with this outcome should allow animals to experience the unpleasurable affective response elicited by it and assign a low incentive value to it. Before exposure to the other outcome, however, we injected the animals with a saline solution in which a small quantity of the anti-emetic ondansetron was dissolved.

Animals were injected with ondansetron because this drug is known to reduce the effectiveness of LiCl in inducing illness, most likely by blocking serotonergic receptors in the brain stem that are usually activated by the viscerogenic effects of LiCl. In addition, in a previous study we found that ondansetron blocked the conditioning of a taste aversion when administered before the injection of LiCl but also attenuated the expression of that aversion when administered before a test of the animals' willingness to eat a previously poisoned food (Balleine *et al.* 1995*b*—experiment 1). This effect of an anti-emetic on taste-aversion learning suggests that the affective response elicited when animals recontact a poisoned taste is, at least in part, associated with illness, such as a feeling of nausea. We predicted, therefore, that re-exposure to an outcome under ondansetron would attenuate the affective response induced by consummatory contact with a previously poisoned outcome and so block the attribution of a low incentive value to that outcome. After this re-exposure phase was completed, animals were given a choice extinction test on the levers and chains. If ondansetron blocks the attribution of a low incentive value to an outcome contacted after it has been paired with illness, we predicted that animals would perform more of the action that, in training, delivered the outcome re-exposed under ondansetron than the other action. That is exactly what we found (Balleine *et al.* 1995*b* —experiment 2).

Table 4.1 Design of experiments assessing the effects of attenuating the affective response elicited by consummatory contact with a poisoned taste using the anti-emetic ondansetron

Training	Re-exposure	Test
Experiment 1		
$(A_1 \rightarrow O_1; A_2 \rightarrow O_2) + LiCl$	OND: O_1 ; SAL: O_2	A_1 vs A_2
Experimental 2		OND
$(A_1 \rightarrow O_1; A_2 \rightarrow O_2) + LiCl$	OND: O_1 ; SAL: O_2	A_1 vs A_2
		SAL

In experiment 1 (top panel) animals were trained to perform two actions, A1 and A2 (lever pressing and chain pulling), earning different outcomes, O1 and O2 (sucrose and saline). Animals were then allowed consummatory contact with both outcomes in separate sessions, one after an injection of ondansetron (OND) and the other after an injection of saline (SAL), before being given a choice extinction test on the levers and chains (A_1 vs A_2). Experiment 2 (lower panel) differs from experiment 1 only in that animals were tested after an injection of either OND or SAL.

We were now in a position to test our main prediction, the design of which is presented in the lower panel of Table 4.1. This experiment was conducted to test directly whether test performance in the previous experiment was mediated by the assignment of an abstract value to the outcomes during the re-exposure phase or by some other means of transfer, such as the animal re-experiencing its affective reactions to the outcomes during the test phase. If, for example, when faced with pressing the lever the rat is reminded of the outcome of that action and re-experiences its affective response to that outcome during the test (it, say, feels ill or nauseous), then this may reduce lever pressing and explain the results of our previous experiment without referring to any change in value assigned during re-exposure. This alternative account would still regard the re-exposure phase as determining test performance but test performance could hardly be attributed solely to the assignment of an abstract value derived from the affective response experienced during the re-exposure phase.

To assess this alternative, an experiment was conducted exactly as that previously described, except that the test conditions were changed (see Table 4.1). Half of the animals were tested as in the previous experiment except that they were injected with saline before the test. We did not expect that this injection would affect performance and predicted that we should find the same pattern of results in this group as in the previous test. The remaining animals were, however, tested after an injection of the anti-emetic ondansetron. The reason this is important is that, if test performance is mediated by the animals' affective reactions during the test when faced with the different actions, then, by blocking the nauseous reaction with ondansetron, we should expect the difference in choice performance to be attenuated. If, however, test performance is entirely mediated by the assignment of an abstract value on the basis of affective experience during the re-exposure phase, then the administration of ondansetron on test should have no impact on test performance. The results of this study are presented in Fig. 4.4.

As was found in the previous experiment, it is clear from the left panel of Fig. 4.4 that animals tested under saline were sensitive to the effects of re-exposure under ondansetron and performed more of the action whose training outcome was re-exposed under ondansetron (O-ond) after the

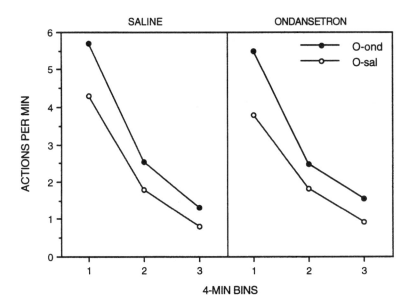

Fig. 4.4 Results of an experiment assessing the effects of attenuating the affective response elicited by consummatory contact with a poisoned taste using the anti-emetic ondansetron, the design of which is presented in the lower panel of Table 4.1. Results are presented separately for the test conducted after an injection of saline (left panel) and the test conducted after an injection of ondansetron (right panel). In each panel, performance of the action whose outcome was re-exposed before the test under ondansetron (O-ond) is presented separately from performance of the action whose outcome was re-exposed under saline (O-sal).

induction of illness relative to the action whose outcome was exposed under saline (O-sal). Far from affecting the size of this difference, however, it is clear from the right panel of Fig. 4.4 that the administration of ondansetron on test had no impact whatever on this effect. Hence, our conclusion from this study is that animals are indeed able to assign an abstract cognitive value to outcomes on the basis of their affective reactions during consummatory contact with that outcome. Importantly, we may conclude that conscious affective states appear to be involved in determining the assignment of an abstract value to a goal but do not appear to be involved in determining the course of action that the animal pursues on the basis of that assignment. Goal value appears, then, to be assigned on the basis of cognitive processing of a conscious affective reaction and integration of this evaluation with beliefs as to the outcome of an action at that level.

These results are consistent with the position that conscious affective processes are necessary for animals to evaluate commodities and so formulate a cognitive evaluation of, or desire for, particular commodities, which may then be described, therefore, as the goals of their actions. From this perspective desires are beliefs about the value of commodities and they are derived from the experienced affective response elicited by consummatory contact with those commodities. This position suggests that, as motivationally pre-potent evaluative beliefs, desires can control actions independently of the conscious expression of that desire, a position that is wholly consistent with Freud's views of the role of the unconscious in psychopathology.

Concluding comments

The argument we have presented here can be summarized as follows. Although basic physiological reactions do a good job of responding to the demands of the environment and in anticipating and preparing the biological system for events that are beneficial or harmful to it, it is clear that there is no sense in which such a system actively controls the external environment. The ability to exert control, actively search for, and evade beneficial and harmful events (rather than passively react to them) confers a clear and undeniable advantage on the agent. But it is equally clear that a great evolutionary change must have occurred for this capacity to emerge, an aspect of which must have been the development of the ability to take cognisance of the relation between environmental events: to form beliefs about 'what leads to what'. But while such a development must have been necessary, it cannot have been sufficient in itself to determine the development of goal-directed action, for information on its own cannot determine a course of action. We cannot, in fact, begin to understand action until we take into consideration the interests of the agent—interests that are usually referred to as desires.

It is usual for us to think of desires, such as to attend the opera or to seek justice for all, as being abstract, culturally determined goals. But it cannot be the case that a desire to eat, drink, breathe, or procreate is so arbitrary. Indeed, we should contend that the desires of any animal must, ultimately, be grounded in its biology. If they were not, if they were grounded in something, an ether, that made no contact whatever with biological needs, any advantage conferred by exerting control over the environment would soon have no material on which to act at all. Desires must, therefore, be grounded in something that the agent cares about. Nevertheless, it is clear that we do not have the requisite perceptual apparatus to allow us direct knowledge of our biological states. So how are the biological states of the agent represented in a way that can be used by a cognitive/intentional system to inform it as to what it should desire?

Our answer to this problem is that the two systems communicate through the medium of consciousness, that the value of particular commodities is grounded in our biology through becoming conscious of affective states. By these states we mean such experiences as pleasure, sweetness, bitterness, anxiety, fear, and so on. In other words, because the agent cannot directly inspect its mechanistic biological states, it can know what to desire only when the effects that particular events have on those states are made explicit through the conscious experience of an affective response. By making the state of the mechanism explicit in this way, the agent can, through experiencing its reactions, assign a particular value to a state of affairs, meaning a desire to obtain or desire to avoid that state of affairs, and can then formulate a course of action on the basis of that assigned value. From this perspective, the most basic function of consciousness is to present a particular physiological state of the organism as an affective response, so allowing the grounding of desires in the biology and the pursuit of things that have biological utility.

This position provides a strong functional approach to the study of consciousness. Furthermore, it suggests that consciousness is the medium that supports the conjunction of a cognitive (or perceptual) representation of a stimulus with the affective response elicited by that stimulus, a conjunction that yields a purely cognitive or abstract desire for that stimulus. From this viewpoint, in line with the inessentialist position, conscious *cognitive* acts are not considered fundamental to the role that consciousness plays in behaviour. A straightforward account can be developed of these conscious acts, however, if they are considered to become conscious through the formation of an association with an affective state; the stream of thought could, for example, become conscious with the ability symbolically to represent events and the environmental contingencies in which those events are embedded as those symbolic representations become associated with affective processes.

Dennett (1991) has argued that any animal that develops a method of implementing a course of action is faced with a primordial problem: 'Now what to do' (p. 177). To solve this problem you need a nervous system, something to control your activities in time and space. And the key to control, Dennett contends, is the ability to track or to anticipate the important features of the environment. He concludes, therefore, that 'all brains are, in essence, anticipation machines'. (p. 177). The aim of Dennett's argument is to develop a theory of the evolution of consciousness, but his account of action very quickly becomes embroiled in more and more complex representational functions. The need to track environmental contingencies, to collect information, becomes, for Dennett, the motivation for all mental acts. As a result, he argued, the real problem facing any sentient creature quickly becomes 'what to think about next' (p. 188).

This kind of question presents Dennett with the motivation to continue his analysis. And it may be that he would contend that our arguments lead us to a similar need for further analysis because the conscious animal, from our perspective, must be continually faced with the problem of *what to feel next.* But that problem admits of a fairly straightforward solution. We feel as our biology needs—we think as we feel, and we do as we think. Only when we start with doing or thinking—and so obscure the motivational source—does the problem of what to do or think arise. If we were truly computational machines, then the problems Dennett is concerned with would be dilemmas indeed. For, without a universal programmes, there would be no reason for the agent to act or think at all. But there is no mystery where our programs come from—they come from our biology. We have argued that the integration of desires with beliefs concerning the way in which these desires may be fulfilled takes place at a cognitive level whatever the goal of those actions may be. Nevertheless, we should not neglect the fact that the power for action comes from our desires whose ultimate source lies in conscious affective reactions derived from the interaction of our (physiologically based) motivational systems with the events or commodities to which they are, innately, tuned to respond.

Careful consideration of this point may lead us to feel that when we cognize, believe, or 'know', we are doing nothing more than pursuing things that satisfy processes in our biology. It may be that there is no strong division to be found in such cognitive states as believing and desiring, that, as Freud would have it, the fundamental unit of our psychological life is 'the wish'. This argument is put forcefully by Henry Miller in *Tropic of Cancer*, where he states:

Still I can't get it out of my mind what a discrepancy there is between ideas and living. A permanent dislocation, though we try to cover the two with a bright awning. And it won't go. Ideas have to be wedded to action; if there is no sex, no vitality in them, there is no action. Ideas cannot exist alone in the vacuum of the mind. Ideas are related to living: liver ideas, kidney ideas, interstitial ideas, etc. If it were only for the sake of an idea Copernicus would have smashed the existent macrocosm and Columbus would have foundered in the Sargasso Sea. The aesthetics of the idea breeds flowerpots and flowerpots you put on the window sill. But if there is no rain or sun of what use putting flowerpots outside the window? (1934, p. 246).

Although the scientific community seems, at present, to be focusing almost exclusively on how to implement conscious cognitive processes in a computational system, we believe that the future of the study of consciousness must reside in furthering our understanding of affective processes and, relatedly, how desires arise from the physiological systems that ultimately determine our evaluation of the goals of our actions. We can do no more than commend the study of these issues to the interested reader.

Acknowledgement

The preparation of this chapter was supported by grants from the National Institute of Mental Health, NIMH grant no. MH 56446, and the European Commission BIOMED 2 programme.

References

Baars, B. (1988). *A cognitive theory of consciousness.* Cambridge, Cambridge University Press.

Baeyens, F., Eelen, P., van den Berg, H., and Crombez, G. (1990). Flavor–flavor and colour–flavor conditioning in humans. *Learning and Motivation,* **21**, 434–55.

Balleine, B. W. (1992). The role of incentive learning in instrumental performance following shifts in primary motivation. *Journal of Experimental Psychology: Animal Behavior Processes,* **18**, 236–50.

Balleine, B. W. and Dickinson, A. (1991). Instrumental performance following reinforcer devaluation depends upon incentive learning. *Quarterly Journal of Experimental Psychology,* **43B**, 279–96.

Balleine, B. W. and Dickinson, A. (1994). The role of cholecystokinin in the motivational control of instrumental action. *Behavioral Neuroscience,* **108**, 590–605.

Balleine, B. W., Ball, J., and Dickinson, A. (1995a). Cholecystokinin attenuates incentive learning in rats. *Behavioral Neuroscience,* **109**, 312–19.

Balleine, B. W., Garner, C., and Dickinson, A. (1995*b*). Instrumental outcome-devaluation is attenuated by ondansetron. *Quarterly Journal of Experimental Psychology*, **48B**, 235–51.

Berridge, K. C. (1991). Modulation of taste affect by hunger, caloric satiety, and sensory-specific satiety in the rat. *Appetite*, **16**, 103–20.

Berridge, K. C., Flynn, F. W., Schulkin, J., and Grill, H. J. (1984). Sodium depletion enhances salt palatability in rats. *Behavioral Neuroscience*, **98**, 652–60.

Berridge, K. C., Grill, H. J., and Norgren, R. (1981). Relation of consummatory responses and preabsorptive insulin release to palatability and learned taste aversions. *Journal of Comparative and Physiological Psychology*, **95**, 363–82.

Borger, R. (1970). Comment on Charles Taylor's 'The explanation of purposive behaviour'. In *Explanation in the behavioural sciences* (ed. R. Borger and F. Cioffi), pp. 80–8. Cambridge University Press, Cambridge.

Brandon, R. (1990). *Adaptation and environment.* Princeton University Press, Princeton.

Cabanac, M. (1971). Physiological role of pleasure. *Science*, **173**, 1103–7.

Cabanac, M. (1990). Taste: the maximization of multidimensional pleasure. In *Taste, experience and feeding* (ed. E. D. Capaldi and T. L. Powley), pp. 28–42. American Psychological Association, Washington.

Cannon, W. B. (1947). *The wisdom of the body.* Kegan Paul, London.

Capaldi, E. D. (1992). Conditioned food preferences. In *The psychology of learning and motivation* (ed. D. L. Medin), pp. 1–33. Academic Press, San Diego.

Chalmers, D. (1996). *The conscious mind: in search of a fundamental theory.* Oxford University Press, Oxford.

Crick, F. and Koch, C. (1990). Toward a neurobiological theory of consciousness. *Seminars in the Neurosciences*, **2**, 263–75.

Davies, M. and Humphreys, G. W. (1993). Introduction. In *Consciousness* (ed. M. Davies and G. W. Humphreys), pp. 1–39. Blackwell, Oxford.

Dennett, D. C. (1991). *Consciousness explained.* Penguin Books, London.

Dickinson, A. and Balleine, B. W. (1993). Actions and responses: the dual psychology of behaviour. In *Spatial representation* (ed. N. Eilan, R. McCarthy, and M. W. Brewer), pp. 277–93. Blackwell, Oxford.

Dickinson, A. and Balleine, B. W. (1994). Motivational control of goal-directed action. *Animal Learning and Behavior*, **22**, 1–18.

Dretske, F. (1995). *Naturalizing the mind.* MIT Press, Cambridge, Mass.

Everitt, B. J. and Stacey, P. (1987). Studies of instrumental behavior with sexual reinforcement in male rats (*Rattus norvegicus*): II. Effects of preoptic area lesions, castration and testosterone. *Journal of Comparative Psychology*, **101**, 407–19.

Flanagan, O. and Polger, T. (1995). Zombies and the function of consciousness. *Journal of Consciousness Studies*, **2**, 313–21.

Frese, M. and Sabini, J. (1985). *Goal directed behavior: the concept of action in psychology.* Lawrence Erlbaum, Hillsdale, N.J.

Freud, S. (1915). The unconscious. In *The Pelican Freud Library*, Vol. 11 (ed. J. Strachey), pp. 167–222. Penguin Books, Harmondsworth.

Garcia, J. (1989). Food for Tolman: cognition and cathexis in concert. In *Aversion, avoidance and anxiety* (ed. T. Archer and L.-G. Nilsson), pp. 45–85. Lawrence Erlbaum, Hillsdale, N.J.

Garcia, J., Ervin, F. R., and Koelling, R. A. (1966). Learning with prolonged delay of reinforcement. *Psychonomic Science*, **5**, 121–2.

Grill, H. J. and Norgren, R. (1978). The taste reactivity test: I. Mimetic responses to gustatory stimuli in neurologically normal rats. *Brain Research*, **143**, 263–79.

Hendersen, R. W. and Graham, J. (1979). Avoidance of heat by rats: effects of thermal context on the rapidity of extinction. *Learning and Motivation*, **10**, 351–63.

Humphrey, N. (1983). *Consciousness regained.* Oxford University Press, Oxford.

Humphrey, N. (1992). *A history of the mind.* Chatto & Windus, London.

Jaynes, J. (1976). *The origins of consciousness in the breakdown of the bicameral mind.* Houghton Mifflin, Boston.

Jerison, H. (1973). *Evolution of the brain and intelligence.* Academic Press, New York.

Lopez, M., Balleine, B. W. and Dickinson, A. (1992). Incentive learning and the motivational control of instrumental performance by thirst. *Animal Learning and Behavior*, **20**, 322–8.

McDougall, W. (1912). *Psychology: the study of behaviour.* Williams and Norgate, London.

McDougall, W. (1925). *An introduction to social psychology* (20th edn). Methuen, London.

Mackie, J. L. (1974). *The cement of the universe.* Oxford University Press, Oxford.

Maze, J. R. (1983). *The meaning of behaviour.* Allen and Unwin, London.

Miller, H. (1934). *The tropic of cancer.* Obelisk Press, Paris.

Pettit, P. (1979). Rationalisation and the art of explaining action. In *Philosophical problems in psychology* (ed. N. Bolton), pp. 3–19. Methuen, London.

Rolls, B. J. (1990). The role of sensory-specific satiety in food intake and food selection. In *Taste, experience and feeding* (ed. E. D. Capaldi and T. L. Powley), pp. 197–209. American Psychological Association, Washington, D.C.

Taylor, C. (1970). The explanation of purposive behaviour. In *Explanation in the behavioural sciences* (ed. R. Borger and F. Cioffi), pp. 49–79. Cambridge University Press, Cambridge.

Woodfield, A. (1976). *Teleology.* Cambridge University Press, Cambridge. T57

CHAPTER 5

The rise of neurogenetic determinism*

STEVEN ROSE

We are now almost through what in the United States has been called the 'Decade of the Brain'. Europe, always slower to move on such matters, is about half way through its own such decade. And we are even further into the massive international $3-billion-odd exercise known as the Human Genome Project—the attempt to map, and subsequently to sequence, the entire DNA alphabet of the human chromosomes. (Identifying just what these DNA strands might do, what the genes might mean, is, as will become clear, a rather different matter, though often elided in popular consciousness.) For a neuroscientist like myself, it is an incredibly exciting time to be in the lab, at the computer, or in the library. New results come flooding in at an almost impossible rate to digest. From the gene sequences of molecular biology to the windows onto the brain produced by new imaging techniques, extraordinary pictures of complexity at all levels, from the chemical through the cellular to the systemic, are emerging. The brain is being seen as a dynamic entity, an ever-shifting sea of electrical and magnetic fluxes, of chemical currents, of growing and retracting cellular connections, of coherent time-locked oscillations that some have even speculated form the basis of conscious experience. But the data have far outstripped theory; rival schools of connectionists (who claim that brain–mind properties can be simulated in the distributed architecture of parallel processing computers) and chaos theorists (who deny a permanent 'seat' to any mind–brain process) strive to make sense of an information overload that almost inhibits meaning.

Just as the Decade of the Brain has produced dramatic advances in information, so it has also generated ever more strident claims that neuroscience is about to 'solve' the brain and in doing so usher in a new era of what José Delgado, an earlier enthusiast for brain surgery to cure violence, once called a 'psychocivilized society'. The emerging synthesis, which I call neurogenetics,

---

* This text first appeared in *Soundings*, Vol. 2 (ed. D. Massey, S. Hall, and M. Rustin), 1996, pp. 53–70. See also, for an extended treatment, my book *Lifelines*, Penguin, Harmondsworth, 1997.

offers the prospect of identifying, ascribing causal power to, and, if appropriate, of modifying genes that affect the brain and behaviour. Neurogenetics claims to be able to answer the question of where, in a world full of individual pain and social disorder, we should look to explain and to change our condition. It is these claims, rather than the excitements of brain theory for its practitioners, which I wish to address here. If the reasons for our distress lie outside ourselves, it is for the social sciences to understand and for politics to try to resolve the problems. If, however, the causes of our pleasures and our pains, our virtues and our vices, lie predominantly within the biological realm, then it is to neurogenetics that we should look for explanation, and to pharmacology and molecular engineering that we should turn for solutions. Social and biological explanations are not necessarily incompatible, but at any time the emphasis given to each seems to depend less on the state of 'objective' scientific knowledge than on the sociopolitical *Zeitgeist*. In the context of rising public concern about levels of violence, an ideology that stresses personal responsibility and denies even the correlation between poverty and ill-health is likely to reject the social in favour of the individual and his or her biological constitution.

Of course this is to simplify, to imply that the world is divided into mutually incommensurable realms of causation in which problems are *either* social *or* biological. This is not my intention; the phenomena of human existence and experience are always and inexorably simultaneously biological *and* social, and an adequate explanation must involve both. Even this may not be enough: both social and biological sciences deal with the world observed as object; personal experience is by definition subjective. To eliminate this personal element from our attempts to understand the world is to fall into the reductive mechanical materialist trap against which both Marx and Engels inveighed. But such unity of subjective and objective may be even harder to achieve than that of the biological and social, and I cannot even begin to approach that issue here. (I tried to talk about it, however inadequately, in my recent book *The making of memory.*[1])

Let us then remain in the world of the objective. Clearly, for any serious scientist to deny the social in favour of the biological or vice versa would be unthinkable; we are all interactionists now. It is by their deeds that one must judge them, however, and in any search for explanation and intervention it is necessary to seek the appropriate level that effectively determines outcomes. Although only the most extreme reductionist would suggest that we should seek the origins of the Bosnian War in deficiencies in serotonin re-uptake in Dr Kaaradzic's brain, and its cure by the mass prescription of Prozac, many of the arguments offered by neurogenetic determinism are not far removed from such extremes. Give the social its due, the claim runs, but in the last analysis the determinants are surely biological. And anyhow, we have some understanding and possibility of intervention into the biological, but rather little into the social.

This is not a new debate; it has recurred in each generation at least since Darwin's day, and most recently in the 1970s and 1980s in the form of the polemical disputes over the explanatory powers of sociobiology.[2–4] What is new is the way in which the mystique of the new genetics is seen as strengthening the reductionist argument. At its simplest, neurogenetic determinism argues a directly causal relationship between gene and behaviour. A man is homosexual because he has a 'gay brain',[5] itself the product of 'gay genes',[6] and a woman is depressed because she has genes 'for' depression.[7] There is violence on the streets because people have 'violent' or 'criminal' genes;[8] people get drunk because they have genes 'for' alcoholism,[9] and there may be genes 'for' homelessness, according to the then editor of the leading US journal *Science*. (The other day I even came across a claim that there might be genes for 'compulsive shopping'; there are clearly no limits to the power of the alphabet soup of DNA.) What isn't due to the genes may be left to biological insults occurring during pregnancy, birth defects, or early childhood accident.

In a social and political environment conducive to such claims, and which has largely despaired of finding social solutions to social problems, these apparently scientific assertions become magnified by press and politicians, and researchers may argue that their more modest claims are traduced beyond their intentions. Such Pilatism, however, is hard to credit when so much effort is put by researchers themselves into what Nelkin has described as 'selling science'.[10] The press releases, put out by the researchers themselves, which surrounded the publication of LeVay's and Hamer *et al.*'s books and papers,[5,6] claiming to have identified 'the' biological cause of male homosexuality and raising a host of alarmist social and ethical speculations, were couched in language that leaves little need for media magnification.

Reductionism

It is my argument that such naive neurogenetic determinism is based on a faulty reductive sequence by which complex social processes are regarded as 'caused' by, 'explained by', or 'nothing but' the workings out of biological programmes based in the brain or the genes. This reductive sequence runs through steps that include: reification; arbitrary agglomeration; improper quantification; belief in statistical 'normality'; spurious localization; misplaced causality; and dichotomous partitioning between genetic and environmental causes. The core issue is reducibility, which, as Peter Medawar once remarked, comes not as second but as first nature to natural scientists. Thus when Karl Popper, giving the first Medaware lecture to the Royal Society in 1986, offered eight terse reasons why biology was irreducible to physics (of which the fourth was that 'biochemistry cannot be reduced to chemistry'), he incurred the wrath of the distinguished assembly, moving the Nobel Prize-winning crystallographer Max Perutz into a vigorous response. Perutz's life's work, after all, had been to demonstrate the relevance of chemistry to biology, and a couple of weeks later he published a reply to Popper, basing his case on the way in which the molecular structure of haemoglobins varied amongst species depending on their environment. Contrast, for instance, the haemoglobin of a mammal living at relatively low altitudes, such as a camel, with that of a related species, the llama, which lives at high altitude in the Andes, where the air is much thinner; the demands on the oxygen-carrying capacity of the blood in the two mammals therefore differ. The structures are subtly different, in each case better fitting the conditions in which its owners live. Is this not clear evidence that human physiology and biochemistry not merely depend upon, but are reducible to, the chemistry of their component molecules? Game set and match to Perutz?[11]

I think not, but will not here argue that case in more depth. Suffice it to say that no amount of analysis, however detailed, of the molecular structure of haemoglobin can lead to an understanding of the function of that molecule as an oxygen carrier in a living animal; in other words its *meaning* for the system of which it is a part. This is not, of course, at all to deny the power of reductionist analysis as part of our attempt to understand complex systems, nor does it reflect on reductionism as a methodology by which to experiment—there is essentially no other way to work. And it says nothing about abstract philosophical concerns with theory reduction. I am concerned here simply with the efforts to attribute causal explanation of complex social affairs through appeals to neurotransmitter metabolism, brain structures, and genes. It is not necessary to enter into a full-blown defence of irreducibility to identify the flaws in the claims of neurogenetic determinism to explain complex social phenomena. Such phenomena are of their essence historically contingent and framed by meanings that the reductive process loses as surely as the information content of the page on which these words are printed is absent from a chemical study of the paper and ink comprising it. And the issue at stake is not the formal philosophical one, but the question of the appropriate level of organization of matter at which to seek causally effective determinants of the behaviour of individuals and societies.

The US Violence Initiative

Let me take a specific example of considerable current concern: the explanation and treatment of the wave of violence that seems to be spreading through the societies of the industrialized world. The debate about the causes of violence long predates the current furore; only the language in which it is cloaked changes. Two decades ago the focus was not genes but chromosomes, when it was claimed that there was a higher than expected prevalence of men carrying an extra Y chromosome amongst those incarcerated for violent crime. And at the turn of the twentieth century, for the followers of Lombroso, it was physiognomy rather than genes that predicted criminality. Before the time of modern science, it was simpler still; it was sufficient to invoke original sin, or predestination. Even if the extra Y has now gone the way of physiognomy and sin, predestination (albeit now spoken in a medicalized hush rather than a hell-fire rant) still lies at the heart of the argument.

Hitherto, in Britain at least, the focus of explanation has been on personal life history; the impoverished rearing practices of single mothers or the laxly disciplined schooling of the 1960s. But in the US even this explanation is being discarded in favour of a return to original biological sin; the fault, we are told, lies in our (or rather their) genes. The argument was put most clearly in 1992 by the then Director of the National Institutes for Mental Health, Frederick Goodwin, in his proposed Federal Violence Initiative. Noting that violence was concentrated in the US inner cities, and especially amongst blacks, who have, he argued, inherited a cocktail of genetic predispositions, to diabetes, to high blood pressure, and to violent crime, he argued for a research programme to identify some 100 000 inner city children and to investigate the genetic or congenital factors that predispose them to such violent and antisocial behaviour. A few years previously, the psychologist Richard Herrnstein coauthored with James Q. Wilson *Crime and human nature*—in many ways the forerunner of Herrnstein's more recent coauthorship of *The bell curve*—which equally focused on the proposition that violent crime in the US is the prerogative of the poor and black and that its origins lie in 'failures' in their biological constitution.

Now there are many obvious objections to such a proposal. Some point to the fact that these discussions always seem to focus on working-class crime; no one seems to study the heritability of the tendency to commit business fraud, or the biochemical correlates of wife-beating amongst middle-class men. Others worry about the complex and sometimes contradictory web of meanings involved in the very concept of violence. On the one hand the identical act, of a man picking up a gun and shooting another at close range, if sanctioned by the State in times of war, becomes an act of heroism worthy of a medal, whereas if it is carried out in the midst of a drugs deal in a Manchester pub it is a crime punishable by a long term of imprisonment. On the other hand, all sorts of different acts are lumped together—Cantona's attack on an abusive football fan, fights between demonstrators and police, the Russian bombing of Grozny—merge, as if one word, violence, fits them all and that hence their underlying causes are the same.

Goodwin's proposal led to charges of racism, and he has subsequently left the NIMH, but a modified version of the Initiative targeted on specific inner-city areas, such as Chicago, is up and running. A conference based on Goodwin's premises, blocked in the US, was held under the prestigious auspices of the CIBA Foundation in London in January of 1995, and the postponed conference in the US eventually got under way—in the face of pickets alleging racism, in September. Not surprisingly, psychologists and psychiatrists, geneticists and molecular biologists have looked longingly at this particular pork barrel. In 1994, I was telephoned urgently by a well-known California-based therapist, just off to Washington to present a proposal to study biochemical and immunological 'markers' in 'violent, incarcerated criminals'. Would I collaborate with him, he asked, in analysing serotonin levels in fluids derived from spinal taps? Serotonin is a neurotransmitter whose metabolism is affected by several well-known drugs, including, as it happens, the now-notorious Prozac. To say nothing of the ethics of per-forming this type of operation on—literally—captive population, the thought that such a study might provide a causal explanation for the endemic viol-ence of US society is just the sort of simple-mindedness that the Violence Initiative fosters.

Amongst child psychologists the key word has become 'temperament'. This nebulous property is, they claim, to a significant degree heritable. Jerome Kagan suggests that some 10 per cent of the infants he studied show, from a very early age, a tendency to shyness, which in later life expresses itself as aggression. To bolster this deterministic argument, he reported finding an analogous pattern of behaviour in kittens that grow into aggressive cats. Adrian Raine and his colleagues studied a cohort of Danish males, now aged in their mid-20s, and found that children with birth complications, products of an unwanted pregnancy, and failed abortion and who are institutionalized during the first year of life commit a disproportion number of violent crimes (murder, rape, armed robbery), and concluded that 'biological factors play some role in violent behaviour—and the role is not trivial'.[12] That children with such a desperate history become damaged and even criminal adults is an observation that would scarcely surprise even the most socially deterministic criminologist; the inclusion criteria for

Raine's sample are likely to cluster with many other impoverished aspects of the growing child's life history. Most, however, would probably regard Raine's conclusion as a leap of faith justified only by a commitment to biologistic thinking.

No biologist could doubt the premise that individual differences in genes and during development help shape a person's actions and distinguish how one person behaves in a given context from how another behaves, nor that a study of the mechanisms involved in these developmental processes is of great scientific interest. But that is neither the reason why nor the way in which 'violence research' is currently being conducted. Rather, it is framed within a determinist paradigm that seeks the causes of social problems in individual biology, and it is fostered by a political philosophy which rejoices in the privileges that come with inequalities in wealth and power and rejects steps to diminish them. The rate of violent crime and of incarceration is higher in the US than in any other industrialized country. Can it really be the case that there is something unique about the genotype of the US population that so dramatically predisposes it to violence? Furthermore, rates of violence are not static; in both the US and the UK, violent crime has markedly increased in recent years—in the US the death rate amongst young males increased 154 per cent between 1985 and 1994. Such fluctuations between and within societies are quite incompatible with any genetic explanation.

The reductionist cascade

What this account demonstrates is the first two steps in a reductive cascade that characterizes all such determinist thinking: reification and arbitrary agglomeration. *Reification* converts a dynamic process into a static phenomenon. Thus violence, rather than describing an action/activity between persons, or even a person and the natural world, becomes instead a 'character', *aggression*, a thing that can be abstracted from the dynamically interactive system in which it appears and studied in isolation. The same process occurs with 'intelligence', 'altruism', 'homosexuality', and so on. Yet if the activity described by the term violence can be expressed only in an inter-action between individuals, then to reify the process is to lose its meaning.

Arbitrary agglomeration carries reification a step further, lumping together many different reified interactions as all exemplars of the one thing Thus aggression becomes a portmanteau term within which all the many types of event and process catalogued above can be linked, all become manifestations of some unitary underlying property of the individuals, so that identical biological mechanisms are involved in, or even cause, each. Take, for example, the descriptions offered in a widely cited paper by Hans Brunner and his colleagues associating a point mutation in the gene that codes for a particular brain enzyme concerned with neurotransmission with 'abnormal behaviour'.[13] The 'behavioural phenotype' in eight males in this family is described as including 'aggressive outbursts, arson, attempted rape and exhibitionism', activities carried out by subjects 'living in different parts of the country at different times' across three generations. Can such widely differing types of behaviour, described so baldly as to isolate them from social context, appropriately be subsumed under the single heading of aggression? it is unlikely that such an assertion, if made in the context of a study of non-human animal behaviour, would pass muster—certainly if I made comparably crude generalizations on the basis of such sparse data in my study of memory in day-old chicks, the paper would rightly be rejected out of hand. Yet the claims the Brunner paper make have become part of the arsenal of argument used, for example, by the Federal Violence Initiative.

Improper quantification argues that reified and agglomerated characters can be given numerical value. If a person is violence, or intelligent, one can ask how violent, how intelligent, by comparison with other people. IQ is one well-known example, but the quantification of aggression is also revealing, for it illustrates another feature of the reductionist cascade that leads to neurogenetic determinism, the use of animal model. Place an unfamiliar mouse into a cage occupied by a rat, and often the rat will eventually kill the mouse. The time taken for the rat to perform this act is taken as a surrogate for the rat's aggression; some rats will kill quickly, others slowly, or even not at all. The rat that kills in 30 seconds is on this scale twice as aggressive as the rat that takes a minute. Such a measure, dignified as muricidal behaviour, serves as a quantitative index for the study of aggression, ignoring the many other aspects of the rat–mouse interaction—for instance, the dimensions, shape, and degree of familiarity of the cage environment to the participants in the muricidal interaction, whether there are opportunities for retreat or escape, and the prior history of interactions between the pair. And just as time to kill becomes a surrogate for a measure of aggression, so this behaviour in the rat is transmogrified into drive-by gangs shooting up a district in Los Angeles. Genes that affect the muricidal interaction are claimed to have their homologues in humans and, therefore, be explanatory factors here too.

In an entirely trivial sense this could be true. A genetic defect that leads to blindness in rats may have its homologue in humans, and blind humans are, one would assume, less likely to pick up a gun and fire than sighted ones. But this is not what the determinists mean when they make their causal claims for a specific genetic origin for violence.

Belief in statistical normality assumes that in any given population the distribution of such behavioural scores takes a Gaussian form, the bell-shaped curve. The best-known example is IQ, the tests for which successive generations of psychometricians refined and remoulded until it was made to fit (almost) the approved statistical distribution—a manipulation exploited to the hilt in *The bell curve*.[14] But the assumption that the entire population can be distributed along a single dimension to which a single numerical value can be ascribed is to confuse a statistical manipulation for biological phenomenon. There is no biological necessity for such a unidimensional distribution (even for continuously varying genetic traits), nor for one in which the population shows such a convenient spread. (It is perfectly possible to set examinations in which virtually everyone scores 100 per cent; the British university penchant for 10 per cent firsts, 10 per cent thirds, and 10 per cent fails, with everyone else comfortably in the middle, is a convention, not a law of nature.) Yet the power of this reified statistic should not be under-estimated. It conveniently conflates two different concepts of 'normality', implying that to lie outside the permitted range around the norm is to be in some way abnormal, not merely statistically but in the sense that ascribes normative values. Thus homosexuality is abnormal in that only a small percentage of the population are gay or lesbian, and it has been at least until recently normatively unacceptable both legally and religiously. When Herrnstein and Murray called their book *The bell curve* they played precisely into these multiple meanings or reified normality.

Having reified processes into things and arbitrarily quantified them, the reified object ceases to be a property even of the individual, but instead becomes that of a part of the person—this is *spurious localization*. So the penchant for speaking of, for example, schizophrenic brains, genes, or even urine, rather than of brains, genes, or urine derived from a person diagnosed as suffering from schizophrenia. Of course, everyone ought to know (and does, at least on Sundays) that this is a shorthand, but the resonance of 'gay brains' or 'gay genes' does more than merely sell books for their scientific authors; it both reflects and endorses the modes of thought and explanation that constitute neurogenetic determinism, because it disarticulates the complex properties of individuals into isolated and localized lumps of biology, permitting neuroanatomical debates to range over whether gayness is embedded in one or other hypothalamic region or, alternatively, a differently shaped corpus callosum in the brain. Aggression is 'located' in the limbic system, probably the amygdala. In the 1970s US psychosurgeons proposed to treat inner-city violence by amygdalectomizing ghetto militants.[15] Things are a little more sophisticated today; a localization in the brain can also take the form of some chemical imbalance, probably of neurotransmitters, so aggression is now 'caused' by a disorder of serotonin-reuptake mechanisms, and drugs rather than the knife becomes the approved approach. Raine claims to be able to detect reductions in the neural activity of the frontal cortex in 'murderers' as opposed to 'normal' individuals by means of brain scans, and hence to be able to predict 'with 80% certainty' from this biological measure the likelihood of a person being a violent killer.[12] It is not clear what such a measure might show in the brain of a Saddam Hussain, a Ratko Mladic, or a Stormin' Norman. Presumably these would feature amongst Dr Raine's 'normals'.

It is at this point that neurogenetic determinism introduces its misplaced sense of causality. It is probable that during aggressive encounters people show dramatic changes in, for instance, hormones, neurotransmitters, and neurophysiological responses, all of which can be affected by drug treatments. People whose life history includes many such encounters are likely to show lasting differences in a variety of brain and body markers. But to describe such changes as if they were the causes of particular behaviours is to mistake correlation or even consequence for cause. This issue has dogged interpretation of the biochemical and brain correlates of psychiatric disorders for decades, yet it still continues. When one has a cold, one's nose runs. But the nasal mucus is a consequence and not a cause of the infection. When one has a toothache it may be sensible to alleviate the pain by taking aspirin, but the cause of the toothache is not too little aspirin in the brain. Such fallacies are, however, an almost inevitable consequence of the processes of reification and agglomeration, for it there is one single thing called, for instance, alcoholism, then it becomes appropriate to seek a single causative agent; complexity is hard to deal with within the neurogenetic agenda.

Beyond either/or

Several single gene defects are known to lead to drastic dysfunctions of mind and brain. Huntington's disease, with its seemingly inevitable progress towards neurological collapse in middle age, is the classic example, but there are many others: Lesch–Nyhan syndrome, Tay–Sachs disease; the list of rare but devastating conditions is long. But neurogenetics consistently overstates its case, moving seamlessly from single to many genes, from genes with predictable consequences in virtually all known environments to genes with small or highly variable effects, whose norm of reaction extends so far as prevent any claims to predictability. In only a few per cent of all Alzheimer's cases is there a clear genetic involvement, while evidence identifying gene markers for manic depression and schizophrenia have been advanced with extensive publicity, and then quietly withdrawn. At best, the hunt for the genes 'for' these conditions may be able to identify anomalous cases in which the genetic effect is to mimic a more widespread phenotypic condition. (Geneticists speak of phenocopies to emphasize the primacy of their genetic explanations in such cases; I have proposed the term 'genocopy' to help geneticists appreciate their more limited contribution.) What both concepts emphasize is the extent to which multiple pathways may lead to a final common biochemical or behavioural endpoint. What both mask is the possibility that the endpoints may not be in every sense identical. Some diagnosed depressions are ameliorated by one type of drug, some by another, and these distinctions have even been made the basis of diagnosis, in which the pharmacological response rather than the clinical syndrome is made the basis for defining the disorder from which the individual suffers—once again insisting on the primacy of the biochemical over the behavioural, the biological over the social.

There are four main negative consequences of such determinism. The first is limited to the study of biology itself; the damage it does to conceptualizing living processes. The primary given to genetic causes fosters, even amongst researchers whose day-to-day practice ought to convince them otherwise, a linear view of living processes, in which the key to life itself lies in the one-dimensional string of nucleotide bases of DNA, the mythopoeic genome. Witness the metaphors with which molecular geneticists speak of the goals of the Human Genome Project—the 'Holy Grail', the 'Code of Codes', 'the Book of Life', 'genes for' particular conditions, as if the entire four dimensions of an organism—three of space and one of developmental and life history—can be read off from this linear code, like a telephone directory.

Few popular writers are more guilty of this than Richard Dawkins. From his first book *The selfish gene* to the most recent, *Climbing Mount Improbable*, he has shown a rhetorical gift for making plausible the gene's eye view not merely of the individual but of the entire living world. Genes aren't selfish—this term can sensibly be applied only to an organism, not a part thereof—but whereas when back in the 1970s I wrote *The conscious brain* the elision in the title was a deliberate act of paradox, Dawkins seems really to believe that this is the way the world really is. Practising evolutionary and population geneticists and molecular biologists recognize that there are multiple mechanisms at play in the processes whereby species evolve or go extinct, and by which new species are formed. They include, of course, classical natural selection and possibly sexual selection, but they also may include genetic drift, molecular drive, founder effects, and many others. But it remains the case that for Dawkins, as for Darwin, explaining how natural selection *per se* generates new species is a great deal harder than explaining how it succeeds in enabling already existing species to evolve so as to get better at doing their own specific things. Furthermore, the selective processes on which Darwinian evolution depends operate at many levels, that of the gene, the genome, the organism, even the population. For Dawkins only the gene, the primary DNA 'replicator', counts. It is a version of what in the 1930s was called 'bean bag genetics' in which each gene is seen as an individual discrete unit. The history of population genetics since then in the hands of Sewall Wright, Theodosius Dobzhansky, and others has been to transcend such simplicities, but the popularizing schemata of Dawkins cannot deal with complexity. This is why he continually reduces organisms to genes and genes to text—to *information*. 'Life', he writes, 'is just bytes and bytes and bytes of digital information.'[16] He may think that he is nothing but a rather primitively designed computer, but I doubt that the rest of us see ourselves that way. 'Information' is not what life is about; it is about 'meaning', and the one is not even formally reducible to the other.

To clarify what I am getting at, consider the phenomenon that I study myself, memory. Memories are believed to be encoded in the brain in some manner based on the establishment of connections between nerve cells such as to create potential novel circuits—a little crude as a description but it will do for the present. Computer modellers from amongst the Artificial Intelligentsia have had a field day producing wiring diagrams of how such circuits could be created, and an entire theoretical universe has grown up over the past decade, called connectionism, which claims to be able to model memories this way. At a recent workshop on memory, an enthusiastic Oxford-based modeller claim to be able to calculate that in primates a particular brain structure, the hippocampus, could encode precisely 36 000 memories. For him, a memory was synonymous with a particular bit of information. How many bits of information do I need to remember the peculiar sadness of the long September shadows, the colour of my son's hair, or even what I had for last night's dinner? These are simply not calculable in terms of information theory. Yet they are central to life—and I suspect they are for the Oxford popularizer Dawkins as much as the Oxford connectionist. Neither they, nor my personal and intellectual engagement with them, are simply bytes and bytes and bytes.

What do genes do?

So what should one be talking about when one discusses genes? The shorthand phrase of a gene 'for' a condition is profoundly misleading—after all, there aren't really even genes 'for' blue or brown eyes, let alone such complex and historically and socially shaped features of human existence as sexual desire or urban guerillas. The cellular developmental and enzymatic route that results in the manufacture of particular pigments involves many thousand genes; the route to the behavioural manifestations we call desire clearly involves genes, but cannot sensibly be regarded as embodied in them. What there are, of course, are differences between genomes (that is, the entire ensemble of genes that any organism possesses). Thus in any particular genome, the absence of a particular gene may result in the emergence of differences in eye colour. The biologist looking at the effects of particular genetic mutations or deletions studies the functioning of the system in the absence or malfunction of a particular gene. Furthermore, the system is not a passive responder to absence or malfunction, but seeks, by means of developmental plasticity, to compensate for any deficit.

A considerable disservice was done to biology by historical chance. In the early years 1900s two separate subdisciplines emerged: genetics, which essentially asks questions about the origins of *differences* between organisms, and developmental biology, which asks questions about the processes that ensure *similarity*. The careless language of DNA and molecular genetics serves to widen this gap rather than help bridge it so as to open the route towards the synthetic biology that we so badly need. As is well known, chimpanzees and humans share upwards of 98 per cent of their DNA, yet no one would confuse the two phenotypes. We have no idea at present about the developmental rules that lead in one case to the chimp, in the other to the human, but this—surely one of the great unsolved riddles of biology—seems a matter of indifference to most molecularly oriented geneticists.

The other consequences of determinism reach out beyond theory. If the homeless or depressed are so because of a flaw in their biology, their condition cannot be the fault of society, albeit a humane society will attempt, pharmacologically or otherwise, to alleviate their distress. This victim-blaming in its turn generates a sort of fatalism amongst those it stigmatizes: it is not our fault, the problem lies in our biology. Such fatalism can bring its own relief, for less stigma attaches to being the carrier or transmitter of deficient genes than to having been morally responsible. It is striking that leading gay activists in the US have embraced the gay brains/gay genes explanation for their sexual orientation on the explicit grounds that they can no longer be held morally culpable for a 'natural' state, nor can they be seen as dangerously likely to infect others with their 'perverse' tastes.

The final consequence is the subversion of scarce resources. Funding for research and treatment becomes misfocused. The orientation of research funds in Russia towards the molecular biology and genetics of alcoholism is one good example, albeit any rational attempt to explain the prevalence of vodka-sodden drunks on the streets of Moscow would not immediately seize on the peculiar genetics of the Russian population as its starting point. Similarly the Violence Initiative, directed towards seeking the origins of violence in American society in terms of the genotypes of blacks and poor inner-city whites, problems of 'temperament' in toddlers, and deficiencies in serotonin-reuptake mechanisms in incarcerated criminals is clearly going to keep a generation of psychologists, neuropharmacologists, and behaviour geneticists in research funds for a good few years to come. One of the keys to success in science is to identify the appropriate level of analysis at which to seek the determinants of complex phenomena. Yet when the differentials between rich and poor are so great and widening, where the potential rewards of violence may be so great (and if large enough can even be socially sanctioned), especially where, as in the US, there are said to be more than 280 million handguns in private ownership, to look to biology to provide a determining explanation of what is going on is an expensive and foolish diversion.

Even in less-dramatic instances, emphasis on genetic explanations and molecularly oriented research prevents researchers from seeing and studying the obvious. The almost universal conviction amongst biological psychiatrists that schizophrenia is a genetic disorder means that they are unable to respond to the suggestive epidemiological evidence that the diagnosis if schizophrenia in the children of black–white relationships in Britain is severalfold higher than that of either of the parental populations.[17] No genetic model fits this finding as well as an explanation in terms of the racism of the society in which these children grow up. Yet it is well known that with a little ingenuity any phenotypic distribution can be explained genetically, granted appropriate assumptions about the incomplete or masked effects of genes (technically known as partial penetrance and incomplete dominance). It is not hard for a behavioural geneticist to offer as an alternative that the data could be accounted for by assortative mating—that is, that you must be mad to begin with to have a relationship with someone of a different colour from yourself.

There is no doubt that the dramatic increases in neuroscientific knowledge are changing and enriching our understanding of brain and behaviour. There is equally no doubt that, wisely and appropriately used, the new knowledge offers the potential to diminish the degree of human suffering, at least in relatively wealthy industrialized societies. But until the neurosciences and genetics can be broken out of their reductionist mould and relocated within a more integrated understanding of the relationships between the biological, personal, and social, abandoning their unidirectional view of the causes of human action so as to recognize the appropriate, determining level of explanation for complex phenomena—that is, until we can stop looking for the key under the lamppost because that is where the light is, even though a moment's thought will tell us we lost it a long way further up the road—their potential for good remains limited and for misapplication substantial and disturbing.

References

1. Rose, S. P. R. (1992). *The making of memory.* Bantam, London.
2. Wilson, E. O. (1975). *Sociobiology: the new synthesis.* Harvard University Press, Cambridge, Mass.
3. Rose, S. P. R., Lewontin, R. C., and Kamin, L. J. (1984). *Not in our genes.* Penguin, Harmondsworth.
4. Kitcher, P. (1985). *Vaulting ambition: sociobiology and the quest for human nature.* MIT Press, Cambridge, Mass.
5. Levay, S. (1993). *The sexual brain.* MIT Press, Cambridge, Mass.
6. Hamer, D., Hu, S., Magnuson, V. L., Hu, N., and Pattatucci, A. M. L. (1993). *Science,* **261**, 321–7.
7. Cohen, D. B. (1994). *Out of the blue: depression and human nature.* Norton, New York.
8. Reiss, A. and Roth, J. (1993). *Understanding and preventing violence.* National Academy Press, Washington, D.C.
9. Galanter, M. (ed.) (1993). *Recent developments in alcoholism.* Plenum, New York.
10. Nelkin, D. (1987). *Selling science.* Freeman, San Francisco.
11. Perutz, M. (1986). A new view of Darwinism. *New Scientist,* 2 October, pp. 36–8.
12. Quoted in Moir, A. and Jessel, D. (1995). *A mind to crime.* Michael Joseph.
13. Brunner, H. J., Nelen, M., Breakefield, X. O., Ropers, H. H., and van Oost, B. A. (1993). [**title** ?] *Science,* **262**, 578–80.
14. Herrnstein, R. J. and Murray, C. (1994). *The bell curve: intelligence and class structure in American life.* The Free Press, New York.
15. Mark, V. H. and Ervin, F. R. (1970). *Violence and the brain.* Harper and Row, New York.
16. Dawkins, R. (1995). *River out of Eden,* p. 19. Weidenfeld & Nicolson, London.
17. Harrison, G. (1990). *Schizophrenia Bulletin,* **16**, 663–71.

CHAPTER 6

Artificial intelligence and human identity

W. F. CLOCKSIN

It is inevitable that research into artificial intelligence is influenced by underlying tacit assumptions about human identity. These hidden assumptions and taken-for-granted realities are an implicit normative load that is generated through discourse on topics such as nature versus nurture, the mind–body problem, and rationality. In this chapter I shall attempt to pose challenges to this implicit normative load, and argue that we need to adjust it if we wish to make progress in artificial intelligence (AI) research. My concern here is with those aspects of AI research to do with proposing general 'principles of operation' of thought and action. By 'action' I refer to embodied engagements such as communication and participation in human society. The aim of AI is to understand these principles in clear and unambiguous (that is, computational) terms. These principles may form the foundation of computer implementations of thought and action, and may in some ways contribute to explanations of human thought and behaviour.

The new approach I advocate here attempts to be attentive to social, ecological, and narrative issues. Its main consequences are to reject the idea that the brain is a nothing more than a rational processor of symbolic information, and to reject the idea that thought is a kind of abstract problem-solving with a semantics that can be understood apart from its embodiment. Instead, primacy is given to emotional and mimetic responses that serve to engage the whole organism in the life of the communities in which it participates. The new approach is also marked by an attention to issues in critical theory: the handling of narratives taking many forms is seen as the fundamental and distinctively human activity that provides a basis for intelligent behaviour. Intelligence is seen not as the deployment of capabilities for problem-solving, but as the continual and unfinished engagement with the environment that conditions narratives of the individuals and the group simultaneously over several different timescales. The construction of the identity of the intelligent individual involves the appropriation or taking up of positions within the narratives in which it participates. Thus, the new approach argues that an individual's intelligent behaviour is shaped by the meaning ascribed to experience, by its situation in the social matrix, and by practices of self and of relationship into which its life is recruited. This 'social constructionist' perspective is at variance with the dominant structuralist (behaviour reflects the structure of the mind) and functionalist (behaviour is determined by mental components) perspectives that currently inform AI research.

Carrying out artificial intelligence[1] research depends on what one thinks intelligence is. Even if one starts by accepting that intelligence is difficult or impossible to define, one's research begins and ends by being conditioned by unwritten and unspoken (that is, tacit) assumptions about the nature of human identity: in other words, what we think we are, and how we think we think. Even computer programmers who consider themselves at liberty to use any implementation technique whatever, irrespective of whether the technique bears a resemblance to a human thought process, need to relate to, if not work from, a global specification of some sort for human thought and behaviour. Usually, researchers' specifications draw on assumptions about human identity that are informed by and address folk versions of standard philosophical conundrums such as the nature–nurture debate and the mind–body problem to a greater or lesser degree. A newer metaphor from computer science, the distinction between hardware and software, has been added to these assumptions about human nature. Taken as a whole and rarely questioned, these assumptions become reified into a kind of prevailing folk belief, which critical theorists call a 'totalizing metanarrative'.[2] Here is an attempt to articulate the metanarrative motivating most AI researchers:

> Intelligent behaviour is determined by processing carried out by an individual brain. A human as the rational animal has a brain which is a seat of orderliness in a strange environment. Thinking is an abstract process capable of being defined mathematically without reference to a particular implementation. Thinking is a deductive procedure executed by a symbol processor working on logical principles. Computers and brains are typical examples of processor substrata on which thinking processes can be implemented. Intelligence involves puzzle-solving, and the environment is a source of puzzles to solve. Evolution and learning are seen as generic design processes that produce fitter or improved individuals (rather than as the historically contingent outworking of an ongoing process of ontogenesis, say).

Those who tacitly endorse[3] this metanarrative are unlikely to agree with every detail as stated, but this is not the point. I hold nothing against the idea of metanarratives in general, but I would like to argue that the particular assumptions and metaphors conveyed by this metanarrative are at best inadequate and at worse misleading if we expect them to tell us anything about the nature of human identity, intelligence, and behaviour.[4] Mind/body, nature/nurture, hardware/software are all artificial dichotomies. In psychological terms, even rationality is not a 'gold standard' of human thought processes.[5] It is right and proper that such dichotomies should exist in scientific theories, for to study complex systems it is necessary to divide them into parts. But at the same time it should be remembered that the definition of what constitutes a dividing line may itself be a simplifying but ultimately misleading assumption.

While AI's preoccupation with logical symbol-processing as the basis of intelligence has been criticised by Weizenbaum[6] and by Dreyfus,[7] Winograd and Flores[8] were probably the first to call for a re-examination and a challenge to the rationalistic tradition in its influence on AI research. More recently, Dreyfus and Dreyfus[9] have predicted that the failure of AI research will continue until intelligence ceases to be understood as abstract reason and computers cease to be used as reasoning machines. In particular, McDermott[10] has argued that

> the skimpy progress observed so far is no accident, and in fact it is going to be very difficult to do much better in the future. The reason is that the unspoken premise ... that a lot of reasoning can be analysed as deductive or approximately deductive, is erroneous.

It is neither difficult nor particularly useful to criticize AI on the basis of its failure so far to realize the solution to the metanarrative. A more subtle criticism, however, arises from the wider and more general difficulty that the very success of an AI system—success in a limited form according to criteria set by its designers—can confer bogus reinforcement onto the metanarrative tacitly endorsed by the system's designers. Yet the metanarrative is formed not by empirical data, but rather by a cultural inheritance having its origins in Classical and Enlightenment philosophy and nineteenth-century economic theory. Thus there is a legitimate critique that we AI system designers are attempting to create systems in the image of our own cultural conditioning. This culture, which might be described as a modern European male culture, is a phenomenon that tends to reduce the social world to objects of exchange, calculation, and control. Bearing in mind the unavoidable social context of human development, such a reductive approach may be practical but misleading if our goal is to understand principles that relate to human thought and action.

Consider instead the relational and systemic practices in which the intelligent system participates. For example, computer scientists and ecologists take a particular interest in the behaviour of complex systems of interrelating, interdependent components. I shall use the language of engagement, commitment, and appropriation when discussing complex systems. The reason for this is not to engage in anthropomorphism, but to use these terms metaphorically to emphasize the context and provisionality in which components of a complex system operate. For example, the ecologist might use the phrase 'community assembly' to refer to a quality or richness of integration among constituents of a living system. To the ordinary observer, a patch of grass is simply an array of monocultural vegetation mounted on a relatively flat susbtrate, but to the trained eye it is a community of thousands of species of organisms—plant, animal, and those such as viruses whose precise biological status is disputed—interacting at all levels of physical size and intervals of time. Although plants are not believed to feel a sense of commitment to their engagement with the environment, there is a meta-phorical commitment and dependence in the sense that, for example, if one grass plant is withdrawn (say by death or change of habit), a number of qualities of the whole system change.

I say this not simply to illustrate an effect of non-linear dynamics (such as the well-known 'butterfly changing the climate' example, which has been used by others), but to bring to mind the numerous ways in which the plant is contracted to its environment; for example, in providing shade, in consuming water, in needing a habitat, and in providing habitat for others—in short, as a co-modulator in the environment. The computer scientist works with such interactions, though not biological ones, at the more abstract level of formal protocols, taking an interest not only in the interaction between constituents (the synchronization and exchange of information between components), but also how low-level details influence and are influenced by higher-level behaviours of and demands on the overall system. So integration and interaction are the conceptual tools we need to work with, rather than dichotomies.

As my colleague John Daugman has frequently pointed out, many of the goals of AI research are accomplished within biological nervous systems, but using radically different strategies and architectures. For example, in living neural systems there are no formal or numerical representations, commu-nication channels are stochastic, events are asynchronous, components are unreliable and widely distributed, connectivity does not obey precise blueprints, and 'clocking speeds' are thousands of times slower than in computers. Yet the performance of natural systems in real-time tasks involving perception, learning, and motor control, in unpredictable and perilous environments, remains unrivalled. The grand challenge of artificial intelligence research is to understand why this is so.

It has been observed for many years that it is relatively easy to program computers to accomplish skilled tasks that are difficult for people to do (such as solving huge systems of equations), while at the same time it has proved impossible to design a computer program to do things that people can do easily (such as washing dishes). To AI researchers this distinction is unexpected, puzzling, and a block to further progress. There are no convincing demonstrations of computers carrying out the range of tasks of perception, motor control, and social interaction that even a year-old infant can do easily. One may ask how it is that artificial intelligence research has been impeded or limited. Is it because research is resource-limited? That is, our theories are right, but our computers do not have enough processing power or memory capacity to do the job properly? Or are we complexity-limited: intelligence is something mysterious that emerges when the under-lying process is complex enough that we no longer understand it, and that we happen to be unable to arrange such complexity at the moment? Or is it that we are concept-limited: that AI is not impossible, but that we aren't thinking about it in the right way? Or is the undertaking fundamentally impossible?

I do not wish to argue, as Penrose[11] and Searle[12] appear to do, that AI is impossible. I believe that computers, if programmed in the appropriate way and subjected to the relevant experiences, may actually think. I am optimistic about the prospects for AI research, but I agree with Weizenbaum that 'intelligence manifests itself only relative to specific social and cultural contexts'. For example, the practices of washing dishes or changing babies' diapers or hammering nails or installing automobile windshields can be taught within minutes to people with widely varying intellectual abilities, creativity, and insight. These tasks involve the handling of imprecise quantities of intractable materials such as soap suds and cloth and slippery dishes and glass sheets, all difficult to model mathematically. In these tasks no precise measurement is called for ('squirt about this much soap into the sink'), specifications are incomplete ('then jiggle it round until it fits'), and circular ('it fits when it snaps into place'), yet dextrous perceptual–motor operations and sensitive judgements are used. The fact that apparently easy tasks may conceal a large amount of unconscious sophisticated analysis and processing is not the point. Whether these tasks need 'insight' or 'representations', or 'knowledge' is also not the point. The point is that these tasks are taught by 'showing how' and learned 'by doing', relying on social and mimetic interaction. This interaction has a spiral 'instituting' effect, as those to whom these tasks are taught can carry them out, and teach them in turn.

A number of years ago it was remarked (not by me) at an AI con-ference that the only behaviour a computer can simulate convincingly is autism. This remark traded on the popular (and incorrect) notion of autism as catatonia, was intended as a joke in poor taste, and got a laugh from the audience. But it is nearer the truth than its teller ever could have known. According to DSM-IV[13], autistic disorder is a pervasive developmental disorder that features qualitative impairment in social interaction. These key features—development and socialization—are the starting points for a fresh approach to the problem of AI that attempts to do justice to questions of human identity.

The problem of arbitrary decomposition

AI theories are generally defined in terms of functional modules. The work of AI falls broadly along divisions of sensory modalities, as the AI subject is divided into research areas such as vision, language, planning, and so forth. Take vision as an example. Analysis may start with a digitized image, and proceed through distinct levels of processing that make certain features of the image explicit. The pattern of colours in the image may be segmented into regions of uniform texture, then these regions can form shapes, then the shapes can define the features of a person's face, then the face can be identified as John Smith's face. The problem of recognizing John Smith is thus decomposed into problems of recognizing its component parts at different levels of description. At one level the presence of circular blobs, say, is signalled by detectors or modules sensitive to circular blobs. At a higher (in the direction of the 'whole') level, a module may discriminate John Smith's face from other faces. Many of the strong assumptions of AI and of neuro-science are concerned with the definition of these modules. As it happens, there is neurophysiological evidence—no, there are only ever observations purporting to adduce evidence—for line detectors, coloured shape detectors, texture detectors, facial expression detectors, in primate visual systems. One difficulty of using this body of knowledge in devising principles of operation is caused by the reinforcement of the assumption that because a particular part of the brain responds to a particular stimulus, that the purpose of that part of the brain is to respond to that stimulus.

Both theoretical models and the empirical observations are based upon two kinds of assumptions. The first is that the perception of the whole is determined by the perception of isolated parts. The second is that an organism's perceptual system is organized in terms of the functional decomposition of modules. Both these assumptions further engender the so-called 'binding problem', in which an explanation is sought for why wholes are perceived when isolated parts are detected. These assumptions arise from a radically reductive approach, but as I mentioned earlier, a certain element of reductionism is proper in science, because complex systems are so large it is judicious to divide them into modules that are easier to study. But at the same time the definition of what constitutes a module and the criteria for distinguishing between modules may themselves be simplifying but ultimately misleading assumptions. A faulty form of such reductionism is demonstrated by nineteenth century phrenological theory, in which alleged mental faculties such as sublimity, cautiousness, benevolence, criminality, and destructiveness were localized and identified with particular areas of the brain. Phrenology is a result of what Rose[14] in another context calls 'reification':

> Reification converts a dynamic process into a static phenomenon. Thus violence, rather than describing an action or activity between persons, or even a person and the natural world, becomes instead a 'character'—aggression—a thing that can be abstracted from the dynamically interactive system in which it appears and studied in isolation.

The trouble with phrenology is not that it suggested locality of brain function. Nor should it be faulted only because it identified particular mental faculties thought at the time to exist. The flaw at the root of phrenology is the tacit assumption 'People do X (or have X as a character), so the brain must have a module for X'. Let us call this assumption modular determinism. Consider the application 'we recognize faces, so there must be a face recognition module in the brain'. At our current state of understanding, face recognition is a plausible suggestion and direction for research.[15] But consider something more unlikely: recognizing pink Volkswagens. Nobody would even bother looking for evidence for a module for recognizing pink Volkswagens; no one presumes such a module to exist. To the adherent of modular determinism, this is not a problem: simply decompose X into generic modules that 'mediate' or 'are prior to' or 'project to', or that 'provide input to' X. Thus, the outputs of modules for recognizing 'pinkness' and for identifying certain models of car provide input to a pink Volkswagen detector. One difficulty is to determine how far this decomposition should proceed, and what determines a particular decomposition. Yet, to the adherent, the need for decomposition to generic modules is not seen as a refutation of modular determinism: it simply makes it easier to believe by abstracting or postponing the description of a covert principle of operation that is somehow causally related to X. Theories (or, more accurately, 'hopes') that rely on modular determinism never tell us how much decomposition is necessary. In that sense, modular determinism is unfalsifiable.

Going a step further, the adherent might claim that 'modules for X are formed in brains that have sufficient experience of X', which is a way of further affirming the unfalsifiability of modular determinism, and which also opens the door to further questions concerning modularity of mind versus modularity of brain. The need to reduce X to generic modules (and, further, to posit their construction), ought to provoke suspicion of modular determinism, because it showed that our guess for X, recognizing pink Volkswagens say, was wrong. But then what is to prevent any more plausible guesses for X being wrong? The adherent would claim that the principle of modular determinism can be applied only at certain levels of description: to recognizing faces, but not to recognizing John Smith, say. The difficulty here is that an appeal is being made to abstract levels of description, which are being defined just as arbitrarily as the problem to which they are being applied.

The study of computer-simulated neural networks—connectionism—has encouraged the misapplication of modular determinism by showing that various Xs of interest may be computed by networks with coefficients whose values are set by iterative improvement, a kind of learning process. It encourages the suggestive image of a correspondence between networks of connections for doing X and modules composed of real nerve cells assumed to do something similar to X. In fact, connectionism has not addressed the more important questions of justifying the choice of X, whether brains need to compute the same sort of X, or how X is related to the system of which it is a part. All connectionism has shown is the feasibility of computing X given training samples of its input/output relation.

Connectionism has been heralded[16] as a fresh start in AI research. But, in fact, connectionism has not changed our ideas of what processing is for and what the intelligent computer should do. It has simply promoted different ways to do it: essentially, an implementation technique based on approximating an unknown input/output relationship from a training set of exemplars. The computer is still carrying out one abstract task in isolation, free of the context of its developmental milieu, defined or parameterized in a way judged appropriate by its author. Whether the computer is programmed by explicit specification of a rule set or by iterative improvement of a matrix of numerical weights is not a sufficiently important distinction on which to claim a new approach to AI. It is popularly thought that the next way forward is to reconcile traditional symbolic AI with connectionism; that is, to formulate a synthesis of the two, or to adopt an integrated position that includes other technologies such as logic programming,[17] or a design methodology based on diverse components (Minsky, ref 3). I believe that all these proposals are unsatisfactory for the same reason as the ideas they seek to supplant: they are nothing more than different ways to implement something, rather than proposals about precisely what needs to be implemented motivated by a good reason *why*.

The binding problem, mentioned earlier, is another misapplication of modular determinism. As Searle describes it, 'when we see an object, we have a unified experience of a single object. How does the brain bind all of these different stimuli into a single, unified experience of an object?'[18] Modular determinism says the brain must have a way to do this, and various neuroscientists have looked for it.[19] The difficulty with this idea is that we never actually see single objects. Simply defining the problem in terms of binding serves only to beg the questions of attention and categorization. There is a level of attention and a quality of engagement with what we see, which, together with our motivations and expectations, plays a part in determining our experience. The problem of binding cannot be carved out as an abstract problem, independent of context and category. Again there is the difficulty of creating a solution to the binding problem in the image of what our culture conditions us to experience.

Modular determinism in the form of phrenology gave license for those in the nineteenth century to impose their assumptions about moral behaviour on a map of the brain. Today we must be careful that modular determinism in a more sophisticated guise does not give license arbitrarily to decompose context-bound socially and historically contingent activities into a selection of abstract, isolated, context-free 'problems' that the brain 'solves' just because they seem natural for us as researchers to specify and represent in that form.

Construction and development

Modular determinism has a temporal analogy, namely the idea that systems develop by passing from one distinct and predetermined stage to the next. The idea of stages of development is popularly attributed to Jean Piaget. In fact, although Piaget's followers were obsessed with the notion of distinct stages of development, Piaget admitted that this 'opens the door to arbitrary thinking'[20] and himself attributed the notion of distinct stages to Sigmund Freud.[21]

To consider the way I handle this problem, namely as the ongoing interaction of capability and experience, it must be restated that intelligent behaviour depends on the challenges that arise from being embodied and being expected to participate in social interactions. Every organism is socialized by dealing with—engaging with, participating with—the problems that confront it according to its capabilities. This has a spiralling effect, as capabilities bring about social experience, which in turn modifies—conditions, improves—the capabilities for further social experience. Instead of 'spiralling', words such as circular, interactive, reciprocal, or unfinished might also be used similarly. An organism's standpoint in this engagement is what brings about—or institutes or constructs or establishes—its identity, independent of the extent to which this is partly apprehended by the organism itself.

Because development occurs throughout an organism's life, its identity is never fixed or foreclosed. Although an organism has a unique identity by virtue of its unique experiences and perspectives, it may establish multiple identities simultaneously according to its capabilities and the demand. For example, when we speak or interpret speech, we are never merely listeners or merely speakers: we are both at once. Hirch calls 'this entertaining of two perspectives at once ... the ground of all human intercourse'[22] although I see no reason to limit the number of available perspectives to two. Furthermore, identity can never be understood as a prior given, even if an organism may understand it (that is, endorses a metanarrative from a culture that names it) as such. AI research cannot begin to make progress until this spiral process of social and cultural development, and its relationship to the ontogenesis of the multiplicity of the individual's identity, can be understood in clear and unambiguous (computational) terms.

The spiral of the interaction of capability and experience can also be described as ontogenesis[23] in terms of a spiral of information uptake by an organism during development. Continuing the example of face recognition, it is important to see such functionality not as a static capability that reaches a certain stage of maturity but as continually conditioned by the experience of the system. Consider the primate who, in the course of behaviour, needs to recognize faces: the animal is wholly engaged in feeding or combat or mating (or some combination of the three, say), and is 'playing' that 'game' in a way that depends on the capabilities it has. Moreover, the capabilities it has depends on the kinds of games it plays and its ancestors played. The animal is in development, undergoing an ontogenetic process that involves playing a part in a drama that is larger than the lifetime of the organism itself; that is, it becomes committed to a circular and unfinished engagement, which is sustained by its current state of development brought about partly by the genotype and environment it inherits, and which facilitates its future development and that of its descendants. Undoubtedly, a capacity for recognizing faces is a crucial component of this engagement, and this capacity becomes practicable at a certain time during the animal's life, but this is not in itself a sufficient reason for pinning down a specific kind of face recognition as a functional module. There is even less justification for computational studies to simulate it as a 'prior given' functional module, detached from its context in the life of the animal.

Another way to describe the developmental process is the so-called 'hermeneutic circle' of critical theory. Interpretation and understanding influence each other in a circular way. To understand a text, we need to interpret it. But understanding it may lead us to consider a fresh interpretation. When the text is read in the light of the new interpretation, it may change our understanding of it. This may call for a revised interpretation, and so on. This circle is again better described as a spiral, for one never visits the same locus in hermeneutic phase-space (as it were) twice. There is also no reason to restrict the number of trajectories through the space to one. Present day AI research projects in language understanding are not attentive to the continual and developmental character of the hermeneutic circle. A system that is not committed to the hermeneutic circle cannot be said to understand. It is the capacity for continual coevolution (in this example, of interpretation and understanding), or ontogenesis, that is characteristic of a sustaining system said to understand. AI techniques for 'belief revision' are not relevant in this regard, as they are techniques for maintaining the consistency of a set of propositions as some propositions are removed or new propositions are added. By contrast, ontogenesis may proceed for a system in which consistency in a logical sense is never achievable. Thus there needs to be an emphasis on the provisionality (or historical contingency) of the system, during which it may not be useful or possible to identify discrete stages of development.

Narratives

The new approach to AI research outlined here is built on the idea of a narrative. The narrative or story provides a framework that facilitates the interpretation of experience, for it is through the narratives people have about their own lives and the lives of others that they make sense of their experiences. Narratives are not restricted to the form of a story with its initial complication followed by its denouement, although stories are good examples of narratives. The idea of narrative is not restricted to written texts. Not only do narratives influence the meaning that people give to experience, they also influence which aspects of experience people select for articulation. Narratives provide not merely a reflection or mirror of life, but provide for the shaping and structure of life. Alasdair MacIntyre[24] recognizes several diverse uses of narrative. He argues that human action is narrative in form, that human life has a fundamentally narrative shape, and that people are story-tellers who position their lives and arguments within narrative histories. Communities and traditions are invested with continuity through narrative histories, and epistemological progress is marked by the construction and reconstruction of more adequate narratives. For Jerome Bruner,[25] narrative is one of the modes of cognitive functioning, a way of knowing that provides a distinctive way of ordering experience and constructing reality. The other mode is logical argument, and Bruner notes that while narrative deals with 'the vicissitudes of human intentions' and is built on concern for the human condition, logical arguments are either conclusive or inconclusive.

Thinking is shaped by a play of images and allusions of the subtle and elusive kind that owes more to the imagination of the story-teller than to the rational operations of the logician. Consider first the question of emplotment. Every story has a plot that involves movement from an initial tension to the resolution of that tension, what Aristotle literally called 'binding' and 'loosing' (*Poetics* 18.1–3), respectively. More generally, a narrative may be described as any time-extended behaviour of a system that exhibits one or more sequential tension → release patterns, each of which may recursively embed other tension → release patterns within it. Each tension → release pattern may be termed an episode. I identify narratives at every scale of time-extended behaviour, from the charge → discharge electrochemical activity of nerve cells to the impulse → relax rhythmic control of co-ordinated motor activity, to the expectation → emission of activity that makes a visual or acoustic impact, to the problem → solution of goal-directed behaviour, to the preparation → resolution of musical and poetic cadence, to the structure of songs and dances, building up all the way to people discussing the latest fashion styles and where they're going to eat lunch. My idea of narrative is informed but not restricted by the ideas of the discursive psychology of Harré and Gillett[26] who are mainly concerned with higher cognitive functions and who see the mind as

> embedded in historical, political, cultural, social, and interpersonal contexts. It is not definable in isolation. And to be a psychological being at all, one must be in possession of some minimal repertoire of the cluster of skills necessary to the management of the discourses into which one may from time to time enter.

Thus discourse can be related to emplotment, and the other typical dimensions of narrative, namely setting and character, are related to environment (context) and identity (being), respectively.

The main difference between narratives and other structures such as knowledge representations[27] is that narratives serve to engage the individual with the contingencies of identity and relationships with other individuals. The individual develops or constructs identity by locating itself within the discourse—the stories, narratives—with which it engages. Thus, identity is not simply a matter of building up a store of knowledge or cognitive capabilities, or the assembly of numerous micro-personalities into a smaller number of personae (*pace* Jung), but the taking up of a position within a set of discourses as they are negotiated. Therefore, behaviour involves a certain performance and production of a self, which is the effect of a discourse, and which has a circular quality. The fact that a developing self may claim to represent itself as a fixed truth before the discourse must be seen as a phenomenon emerging from the engagement rather than as an ontological assumption.

This 'taking up of a position' within a narrative may be identified at all levels: from the steady-state behaviour of an electrochemical process in tissues, to the habituation of a tissue in response to a pattern of conditioning stimuli, to the formation of a reflex. At the highest social level it may involve the articulation of or appropriation of or commitment to a particular policy, a system of beliefs and values, a family group, a football team, or a gender role. The ways in which one plays social games is the way in which the identity becomes, to use Butler's list of words, 'established, instituted, circulated and confirmed'.[28] It is important to remember the ontogenetic nature of the process: the taking up of a position within a narrative is not entirely determined by genetics, nor is it entirely determined by culture. All enabling capabilities participate in their own way with the development of the system.

Let me outline some broad implications of understanding intelligence in terms of narrative and the narrative generation process:

1. Narratives, having a linear articulation,[29] can be repeated and imitated. Aristotle defines man as the political and rational animal, but we should also remember his observation that 'man differs from other animals in his greater aptitude for imitation [mimesis]' (*Poetics* 4). I see Mimesis as the fundamental operator over narratives, implicated in the processes of memory, consciousness, and performance as described below. Mimesis is not a process of loss-free copying: rather, it is a capability that culminates in 're-storying', or re-telling, or re-authoring of a story. Mimesis in the sense considered here is also not to be identified with the Romantic notion of art as 'imitating' life.

2. Narratives, being indeterminate and having hierarchically decomposable emplotment, are a convenient form for concealing and revealing sub-narratives. A narrative is extensible both along its length and its hierarchy, for it is always possible to apply questions such as, 'and *then* what happened?', or 'where did he come from?', or 'how did she do that?', or 'why did he do that?', or 'why didn't she say that?' as a means of closing or opening new sub-narratives and filling in or creating gaps.

3. Narrative serves as a memory, not only in a collective and institutional sense of telling stories to each other (memorial), but also as the foundation for the individual's memory. Memory as narrative provides an alternative to the conventional idea of memory as a storehouse of knowledge and experience, and of remembering as a process of retrieval from the storehouse. Instead, memory is understood not as the storage-and-retrieval manipulation of a

network of entities, but rather as a particular practice within the context of a discourse. Memory is an active process in which experience is emplotted (the lived experience is 'recruited' by a prior narrative) and personified (attributed the status of a 'character' in a prior narrative) in ways constrained by the usages and contexts of prior narratives. This 'circular' definition is committed to the idea of memory not as storage and retrieval of facts, but as the continual re-membering of our memories of our memories (*sic*). Memories cannot be decoupled from the narrative structure of remembering, for outside a narrative they have no meaning.

4. Narratives can be edited, and this is particularly relevant to the question of memory and learning. As Platinga puts it, 'our memories are not inert, but undergo a process of editing, whereby they are regularized, rendered more retainable and reshaped with an eye to subsequent circumstances and events'.[30] Learning and therapy are understood as means of editing or rewriting narratives.[31] Both these methods begin with the assumption that people experience problems when the stories of their lives, as they or others have constructed them, do not sufficiently represent their lived experience. Therapy and learning then become processes of storying and re-storying the lives and experiences of people. In this way narrative comes to play a central role in therapy and education.

5. Narratives can be performed. Performance is the means by which narratives are rendered to self and others, and depends on the capacity for mimesis. Performance is more general than reciting a story on a stage: it is the actual means of behaviour. Yet performance, like remembering, does not render a narrative intact. This understanding concurs with Butler[32] in opposing the conventional notion that performativity is the efficacious expression of the human will. An act of performance is always ambiguous and open to interpretation. Each performance is a unique act of creation because it opens a new gap between intention and realization, between presentation and representation.

6. Because mimesis (the mechanism for re-storying, performance, memory) generates non-identical narratives that are open to interpretation, it is a wellspring of perpetual conflict. Conflict leads to rivalry and violence, but also to dialectic and symbolic interaction. Thus the provision of narrative-building tension together with a motivation for social interaction are built in at a fundamental level.

A narrative is a suitable representation for self-reflection. Whether consciousness is considered as a capacity for, or a result of, self-reflection, consciousness is about making explicit in narrative form (telling a story about) the person's interpretation of the narratives with which it believes itself to be engaged. The nature and extent of the belief are also conditioned by the person's appropriation of narratives from the culture, giving rise to 'cultural conditioning'. Consciousness is the story we tell about where (context) we (character) are, and what (plot) we are doing. Who we tell is part of the constructed self that is doing the telling. The existence of what Dennett calls the 'Cartesian theatre'[33] as a setting for this performance therefore should not be surprising. Consciousness as an 'account of ourselves' therefore proves necessary because we are animals who partly construct our accounts through interaction with a social world: we need accounts of ourselves because we are accountable 'as selves' to others, and this exchange of accountability again has a spiral instituting character. The nature and purpose of consciousness as a performative practice therefore cannot be understood in relation only to the individual or to the goings on within an individual's mind–brain. Dennett is therefore wrong to suggest that the idea of the Cartesian theatre has anything to do with Cartesian dualism.

Narratives which emerge at the layer of surface behaviour are particularly important, for all performance is understood within the context of the social group. At a fundamental level narratives are performed as rituals and myths, both of which also have the instituting effect of maintaining a corporate memory. Myths supply a framework of images, in narrative form, in which the needs, values, and fears of the group—in short, an articulation of emotions within a network of social consciousness—can be expressed. Rituals are a demonstration of the way that sensations and actions are integrated as a result of their codevelopment, and are expressed by the articulation of a policy by members of the group. Recalling Weizenbaum's observation that 'intelligence manifests itself only relative to specific social and cultural contexts', I claim that ritual and myth as the performative means of the culture are the prior and necessary components for the manifestation of intelligent behaviour. This implies a view of intelligence which is an inversion of the conventional view taken by artificial intelligence and cognitive science researchers. The conventional view assumes that behaviour arises from basic capabilities for cognition and problem-solving. The emphasis seems to be on the individual mind–brain that generates solutions to problems. The questions of emotion and interaction with others are considered as additional—possibly even unnecessary or irrelevant—complications or side issues. The apparent appeal of the comparison of the human condition with a 'brain in the vat'[34] is particularly telling in this regard. By contrast, my 'upside-down' view sees problem-solving as a practice or policy that has been constructed as a result of a basic capacity for social interaction within a cultural context, which itself is based on fundamental reciprocalities of need and desire. Therefore, further progress in artificial intelligence research needs to be attentive to issues arising from the articulation of practices and policies by a group.

An example

I shall briefly consider an example of a policy articulated by members of a group. The policy is *sacrifice*, defined as, 'one takes a risk so that others may obtain a benefit'. In my 'upside-down' view of cognition, policies such as this are elementary, at the very foundation of cognitive capabilities. They are not—as the conventional view might suppose—the grand and elaborate product of a civilization or the carrying out of religious laws. In the sacrifice mechanism, a victim is selected from within a group and undergoes a risky trial, which may result in exile or death. This problematic pattern of behaviour might be seen by ethnobiologists simply as an extreme form of altruism on the part of the victim, but this is more a description of the phenomenon than an explanation. Altruism is perhaps more accurately associated with self-sacrifice. Instead, the idea of sacrifice I consider here is not self-sacrifice, but a situation in which one member of a group is selected for sacrifice. This is an important distinction because it immediately brings in the complicating factor that we must be concerned not only with an individual's own inclinations, but also with how an outcome may be achieved by the interaction of members of a group who are more-or-less engaged in the same activity. Sociobiological concerns such as whether the members are cooperating, competing, or being deceptive, or are genetically related is not the issue here. The altruism theory seems to be concerned mainly with how the individual deploys its own resources. This is not an adequate explanation for situations such as sacrifice in which each individual may have the capacity to deploy itself as a victim, but only one member of the group actually does so. I consider the sacrifice mechanism as a policy articulated and endorsed by the group, and as such, its formulation as a narrative seems appropriate. The articulation of policies is also not relevant to the debate in evolutionary biology between individual selection and group selection. Such a debate might be concerned more with biological questions about how altruism comes about. By contrast, my concern is how sacrifice can be seen as having a reciprocal constitutive effect on the narrative stock of the group.

Take as a starting point the work of Girard,[35] who explains the need for sacrifice as a way of dealing with rivalry. Many myths involve original murders, in which one rival, often a brother, murders another. Although the murder may temporarily relieve the tension of rivalry, homicide is ultimately counterproductive to the group as it leads to vendetta, and violence may escalate. Murder, resulting from rivalry, is so abhorrent that according to Girard a special mechanism has developed in order to focus the group's attention on a substitute victim. This requires the substitute to be distinguished in some way—for example, in the case of a scapegoat,[36] to be held responsible for some crisis afflicting the group. So as a culture develops, a single sacrificial ritual develops that punishes the scapegoat, thus relieving the tensions that lead to rivalry and violence.

We can put Girard's treatment in terms of a primitive narrative. At its simplest, the tension of rivalry is resolved by the death of the rival. Stories about the primitive narrative circulate and constitute memorial narrative. The sacrifice ritual serves to perform the narrative, with an instituting effect on the group. But I depart from Girard's treatment by associating the origin of sacrifice with the development of the group within the environment, identifying the (individual's) selection advantage of sacrifice within a group involved with risktaking. To survive in the environment, the group needs to take risks. Such risks might be sampling unfamiliar food, or exploring new territory, or engaging with an unfamiliar adversary, or imparting unwelcome news to the group. If every individual is inclined 'to take up a position' to take the risk, or if none is, the survival of the entire group is threatened. This requires members of the group to endorse a policy in which roles or responses are differentiated, and one member of the group is identified to bear the risk on behalf of the group. The risk is that one member might die, but the advantage is that the group might be saved. There are two stable strategies or 'schedules' that will reinforce (reward) this behaviour: (1) to invest the victim with great value as the innocent and heroic risktaker, or (2) to associate (or 'convict') the victim with the 'cause' of the crisis (scapegoating). Both reinforcement schedules are found in the world's myths. The adoption of schedule 1 can provide an incentive to potential volunteers for the role of one who is elevated to high status, and the heroism of the victim may act to confer reflected glory (or vicarious heroism) on the group by recognizing that the heroic victim is 'one like us'. Alternatively, the adoption of schedule 2 permits the group to identify itself as a just group, putting distance between the group and the evil represented by the scapegoat. Either way, the tension of rivalry is dissolved, and additional details of plot, context, and characterization are provided by environmental and cultural particularities.

Conclusion

For further progress in AI research, intelligence should not be seen as abstract puzzle-solving applied by an individual to arbitrarily defined problems presented by an alien environment. It is true that certain isolated episodes of intelligent behaviour can be given an interpretation as puzzle-solving, maximizing of utility, and so forth. We need to accept, however, that there is a close ontogenetic connection between the nervous system and the environment. It is possible to accept that this connection has the effect of individualizing precisely which capabilities and information sources are used by the organism as it develops. This does not mean endorsing a structuralist ontological commitment as to the precise delineations of capability and information source. Indeed, as the controversies over linguistic categories[37] and the cladist versus pheneticist debates[38] show, agreement on such endorsement is well-nigh impossible even for the supposedly straightforward situation of interpreting empirical data.

The AI system of the future will be based on what Darwin called 'the social instincts'. Intelligence—as we understand it—cannot be considered in isolation from an embodied existence that is committed to an ontogenetic engagement with the environment. The body matters, and in this way intelligence might be said to resemble physical functions such as digestion which cannot take place without a body. But intelligence is of a different order, because at its core is the generation and use of images and narratives that serve to circulate within the gap between presentation and representation that is opened up by performance within the social group. Again we return to Weizenbaum's dictum that intelligence manifests itself only relative to a matrix of social and cultural contexts. To this I also stress the historical contingency of such contexts. There is a propinquity of participation in communities that are defined by historically individualized mutualities and reciprocalities of need and desire. The ways we participate with this matrix, and construct identities that take up positions within this matrix, are through policy, contract, and institution. This participation and construction, which make explicit the political imperative of intelligence, are made possible by a variety of means such as memorial, mimesis, ritual, education, and therapy, all of which serve to expose and rewrite narratives. Human identity is thus seen as an historical contingency constituted in part by policies articulated by the social group, rather than as an underlying cognitive nature of the individual brain or as a realisation of ideal goals. This perspective is what Rorty[39] might call 'ironic' and 'political' in the sense that it attempts to make explicit its theoretical commitment to policy, community and institution.

Biology matters, and so does our aim to understand neurophysiological and cognitive phenomena through the use of computational metaphors. But, if intelligence is based on a capacity for narrative in all the ways indicated above, then AI research is engaged with the entirely legitimate activity of telling stories about the story-telling process. So perhaps our discourse in AI should be less about formal logic and more about critical theory. It should be less about building intelligent systems in the image of our own cultural conditioning, and more about understanding the process by which intelligence emerges from constitutive and contractual engagements with cultural and group systems. One implication for future research is the idea of 'narrative architectures', built upon a foundation that is attentive to the way that signals and symbols influence (and are influenced by) the way we make sense of experience, construct our identities, and produce meaning in the world.

Notes and references

1. For an accessible and non-technical introduction to artificial intelligence see *Artificial intelligence: a philosophical introduction* by Jack Copeland (Blackwell, Oxford, 1993). For a technical and comprehensive textbook I recommend *Artificial intelligence: a modern approach* by Stewart Russell and Peter Norvig (Prentice-Hall, New York, 1995).
2. For a popular treatment of metanarrative and the postmodern condition, see: M. Luntley, 1995. *Reason, truth and self: the postmodern reconditioned* (Routledge, London, 1995) and J. R. Middleton and B. J. Walsh, 1995. *Truth is stranger than it used to be: biblical faith in a postmodern age* (SPCK, London, 1995). An accessible sourcebook of readings in critical theory is *Modern criticism and theory: a reader* by D. Lodge (Longman, Harlow, 1988).
3. Using Minsky's terminology, I expect 'neats' to be strong endorsers, and 'scruffies' to endorse less; see M. Minsky 'Logical versus analogical or symbolic versus connectionist or neat versus scruffy', *AI Magazine* **12** (2), 34–51 (1991). The metanarrative as expressed here may not look much like a story, but it does when placed in the context of the scientific quest, with its drama of a problem and the assumptions and hypotheses contributing to its proposed solution. In its more extreme forms, the AI metanarrative concludes with scenarios such as intelligent computers keeping us as pets, or of human minds being 'transferred' into immortal computers, or as computers being the post-human stage of evolution. For examples where such eschatology is apparently taken seriously see H. Moravec, *Mind children: the future of robot and human intelligence* (Harvard University Press, Cambridge, Mass., 1988).
4. The entirely legitimate project of exposing and challenging tacit assumptions and taken-for-granted realities can be described by the much abused term 'deconstructionism'.
5. For examples, see S. Sutherland, *Irrationality: the enemy within* (Penguin, London, 1992); M. Piattelli-Palmarini, *Inevitable illusions: how mistakes of reason rule our minds* (MIT Press, Cambridge, Mass., 1994);N. Dixon, *On the psychology of military incompetence* (Futura, London, 1976); N. Dixon, *Our own worst enemy* (Futura, London, 1987).
6 J. Weizenbaum, *Computer power and human reason* (Penguin, London, 1976).
7. H. L. Dreyfus, *What computers can't do* (MIT Press, Cambridge, Mass., 1972).
8. T. Winograd and F. Flores, *Understanding computers and cognition: a new foundation for design* (Addison-Wesley, 1985).

9. H. L. Dreyfus and S. E. Dreyfus, *Mind over machine* (Macmillan, London, 1986).

10. D. McDermott, 'A critique of pure reason', *Computational Intelligence*, **3**, 151–60 (1987); here, p. 151.

11. R. Penrose, *The emperor's new mind* (Vintage, London, 1990).

12. J. R. Searle, 'Is the brain's mind a computer program', *Scientific American*, **262** (1), 20–5 (1990).

13. American Psychiatric Association, *Diagnostic and statistical manual of mental disorders*, 4th edn (Washington, D.C., 1994).

14. Stephen Rose, 'The rise of neurogenetic determinism', *Nature*, **373**, 380–2 (1995); here, p. 380.

15. See M. H. Johnson and J. Morton, *Biology and cognitive development: the case of face recognition* (Blackwell, Oxford, 1991).

16. See Paul Churchland, *The engine of reason, the seat of the soul* (MIT Press, Cambridge, Mass., 1996).

17. R. Omar, 'Artificial Intelligence through logic?', *AI Communications*, **7** (3/4), 161–74 (1994).

18. J. Searle, 'The mystery of consciousness', *The New York Review* of Books (2 November), p. 64 (1995).

19. See, for example, F. Crick, *The astonishing hypothesis: the scientific search for the soul* (Simon & Schuster, New York, 1995).

20. J. Piaget, *Biology and knowledge* (Edinburgh University Press, Edinburgh, 1971).

21. For a survey of biological and cognitive psychological perspectives on development, see M. H. Johnson, ed., *Brain development and cognition: a reader* (Blackwell, Oxford, 1993).

22. E. D. Hirsch, *The aims of interpretation* (University of Chicago Press, Chicago, 1976), pp. 36–49.

23. S. Oyama, *The ontogeny of information* (Cambridge University Press, Cambridge, 1985).

24. A. MacIntyre, *After virtue: a study in moral theory* (Duckworth, London, 1981).

25. J. Bruner, *Actual minds, possible worlds* (Harvard University Press, Cambridge, Mass., 1986).

26. R. Harré and G. Gillett, *The discursive mind* (Sage Publications, London, 1994), here, pp. 25–6.

27. E. C. Way, *Knowledge representation and metaphor* (Kluwer, Dordrecht, The Netherlands, 1991).

28. J. Butler, 'Imitation and gender insubordination', *Inside/Out*, ed. D. Fuss *et al.* (Routledge, London, 1991), pp. 13–31, here p. 18.

29. This rather technical description does not imply that narratives necessarily have an internal linear representation.

30. T. Platinga, *How memory shapes narratives* (Mellen, Lampeter, 1992), p. 45.

31. For what might be described in my terms of narrative–theoretic formulations of memorial, mimesis, ritual, education, and therapy, see the following respectively: C. M. Elliott, *Memory and salvation* (SPCK, London, 1995); R. Girard, *Violence and the sacred* (Johns Hopkins University Press, Baltimore, 1977); S. Buckland, 'Ritual, bodies and "Cultural Memory"', *Concilium* **1995** (3), 49–56 (1995); D. Epston and M. White, *Experience, contradiction, narrative and imagination* (Dulwich Centre Publications, Adelaide, South Australia, 1992).

32. J. Butler, *Bodies that matter* (Routledge, London, 1993), p. 187.

33. D. C. Dennett, *Consciousness explained* (Penguin, London, 1991).

34. Dennett, *Consciousness explained*, p. 3.
35. R. Girard, *Violence and the sacred* (Johns Hopkins University Press, Baltimore, 1977). Girard's use of the term 'mimesis' is not identical to mine, although there are certain points of contact. Girard, who does not explicitly use the notion of narrative, is concerned with imitative desire and rivalry, while I confine mimesis to its function as a re-storying mechanism.
36. R. Girard, *The scapegoat* (Johns Hopkins University, Baltimore, 1986).
37. G. Lakoff, *Women, fire, and dangerous things* (University of Chicago Press, Chicago, 1987).
38. S. J. Gould, *Hen's teeth and horse's shoes* (Norton, New York, 1983).
39. R. Rorty, *Contingency, irony, and solidarity* (Cambridge University Press, Cambridge, 1989).

CHAPTER 7

Quantum curiosities of psychophysics

JEREMY BUTTERFIELD

1 Introduction

My subject is quantum theory and the mind. These are very disparate topics, so it can seem foolhardy to link them. But I believe there are some genuine connections. More specifically, I believe there are connections between two philosophical aspects: between the problems about interpreting quantum theory, especially the so-called 'measurement problem', and the metaphysics of the relation between mind and matter. In this chapter, I will survey some of these connections. I will emphasize the metaphysics, especially in the first half (sections 2 and 3), and I turn only in the second half (sections 4 and 5) to how quantum theory's measurement problem bears on the metaphysics. (In two complementary papers I emphasize the quantum theoretic issues, and discuss 'no-collapse' solutions to the measurement problem, which I set aside here: Butterfield 1995, 1996.)

I will have two main conclusions. They are both about how some ways of solving the measurement problem (that is, some interpretations of quantum theory) bear on the doctrine of physicalism. This is the view, roughly speaking, that everything is physical: less roughly, that all facts, and *a fortiori* all mental facts, are actually physical facts.

My first conclusion is that although some interpretations of quantum theory, which I will present, are compatible with physicalism (according to various exact definitions), they involve a very different conception of the physical, and of how the physical underpins the mental, from what most people expect—including philosophers and neuroscientists participating in the debate about whether the mind is physical. In short, the interpretation of quantum theory holds some surprises for this debate. Hence my title!

My second conclusion is more specific. It is that one interpretation of quantum theory gives a real-life example of a problem, which metaphysical discussions of physicalism have seen only in the abstract. It is a problem about how exactly to define physicalism. Namely, one attractive and much-discussed strategy for making physicalism precise threatens to be too weak; that is, the precise definition can be satisfied, even while intuitively physicalism is false. So my point will be that one interpretation of quantum theory illustrates this problem.

To establish these conclusions, I first need to summarize some aspects of the contemporary debate about the mind–matter relation, leading up to the much-discussed definition of physicalism. (To anticipate for a moment, using the jargon of philosophy, physicalism is defined as a contingent supervenience thesis.) This summary is in sections 2 and 3.

Section 2 clears the ground. It has three subsections. In these, I argue, first, that there is a prima-facie distinction between mind and matter (2.1); second that some mental states are literally in our heads (2.2); and, third, that even if these mental states are identical with states of our brain, no dubious 'reductionism' follows (2.3).

Section 3 formulates physicalism. It has four subsections. In 3.1, I distinguish physicalism from the logically weaker doctrine of materialism (the doctrine, very roughly, that everything is completely describable by the natural sciences). In 3.2, I distinguish two senses in which it could be true that 'all facts are physical facts'. The first, I call 'reduction'; the second, 'supervenience'. Reduction is logically stronger: that is, it implies supervenience. In 3.3, I spell out the details of these senses—and meet an obstacle. Namely, if we consider only actual facts, physicalism in these two senses is in grave danger of being trivially true, as a matter of mere accident. In 3.4, I describe how to overcome this obstacle: we need to consider not only the actual facts, but also possible facts—how the history of the Universe might have gone. But exactly which possible facts? One much-discussed answer is 'the possible facts that accord with the actual laws of nature'. It is this answer that faces the problem mentioned above. It makes the formulation of physicalism too weak: physicalism comes out true in cases where intuitively the mental is *not* physical, but is just rigidly connected to the physical by laws of nature.

That will conclude my general discussion of physicalism. In section 4 I turn to quantum theory. First, in 4.1, I warn against a naive picture of physical reality as composed of particles like billiard balls moving in a void. This picture has led to bad arguments about physicalism. In particular, I rebut a bad traditional argument *for* physicalism, which is still endorsed by some authors. In 4.2 and 4.3, I describe quantum theory's measurement problem. Roughly speaking, it is this: quantum theory's laws about how the states of objects change over time seem committed to the prediction that macroscopic objects often have no definite positions—nor definite values for other familiar physical quantities like momentum or energy. But this seems manifestly false: tables and chairs surely have definite positions, etc. As it is sometimes put, the macrorealm is definite; or at least we experience the macrorealm as being definite. So, if quantum theory is to account for our experience, it must either secure such definiteness or at least explain the appearance of it.

Section 5 covers one approach to solving this problem. This approach postulates new physical laws that avoid the false prediction. (I set aside another approach, in which the ordinary laws of quantum theory are retained, but the false prediction is avoided by postulating 'extra values' for some physical quantities. Suffice it to say that, as in section 5, some versions of that approach are compatible with physicalism, but others are not: see Butterfield 1995, especially pp. 145f.). I will consider three versions of this approach. The first two versions (5.1) are compatible with physicalism; but they yield my first conclusion. That is, on these proposals, the *way* in which the mental is fixed by the physical would be very different from what most physicalists, unversed in the controversies about quantum theory, would expect. The third version (5.2) yields my second conclusion. Here is an interpretation of quantum theory that illustrates the problem at the end of subsection 3.4: an interpretation that violates the spirit of physicalism, if not its letter.

2 Mind and matter

In this section and the next, I will summarize some aspects of the contemporary debate about the mind–matter relation. I have chosen these aspects with a view to complementing other chapters in this volume, especially those by Boden, Lipton and Searle, which I will try to locate in the debate.

We have countless beliefs about the material (empirical) world in space and time, including our bodies, and countless beliefs about the mental world—that is, the minds and experiences of ourselves and others, including perhaps animals. So here are two subject-matters for our beliefs, each very large and heterogeneous (and they are also subject-matters for our other attitudes: desire, question, etc.). The question, then, arises: what is the relation between these subject-matters? That is our problem, the so-called 'mind–body problem'; though because 'body' has the specific connotation 'animal organism', rather than 'material object', it would clearly be better to call it the 'mind–matter problem'. As stated, the problem is vague: we have no precise notions of subject-matters, and of relations between them. Once we clarify those notions, the problem will no doubt break up into different problems, some scientific and some philosophical.

In the first place, it is not easy to state exactly what these two subject-matters are. Accordingly, therefore, many authors take a sceptical view of the whole debate. So I first address this view.

2.1   Denying the mind–matter distinction

These authors doubt that a precise distinction between the subject-matters can be made; or they doubt that, once it is made, it is scientifically and/or philosophically significant—that it in some way 'carves nature at the joints'. (For example, Midgley argues that materialism and idealism are both half-truths, to be transcended; see Chapter 8.)

This scepticism is often supported by two related lines of argument. First, one can spell out the historical process by which this distinction entered our philosophical culture. The orthodox idea is, of course, that it entered through the mechanical philosophy of Galileo, Hobbes, Descartes, and their successors, who shared a common vision of a mathematical mechanics underlying all material facts, though perhaps not that apparently very disparate arena—the mind. In spelling out this historical process, one discovers, unsurprisingly, that there was less of a common vision than this orthodoxy suggests; and one gets a vivid sense that our contemporary mind–matter distinction is historically moulded, and is not an intellectual necessity (for example, Baker and Morris 1993). Both points tend to undermine one's conviction in the scientific or philosophical importance of that distinction.

Second, one can argue that the development of science since 1700 has not vindicated the alleged common vision. There is an uncontentious point here: the concept of a mathematical physics, as the 'basic science', has changed drastically, above all by giving up the primacy of mechanical concepts.[1]

But this second line of argument also makes a contentious claim: that physics, even as transformed since 1700, is in no sense 'basic'; that on the contrary, modern science is in fact very disunited, with the separate sciences pursuing their own goals, and developing their own theories, entirely regardless of physics (for example, Dupre 1993). Such authors often go on to argue that this disunity seems likely to continue—it is not a temporary predicament. So they propose to drop the binary mind–matter distinction, as an artefact of a certain philosophical legacy; and to replace it by a picture of many subject-matters studied by the many separate sciences. These subject-matters are no doubt connected in various ways; for example, by causation (that is, an event or state of affairs in one subject-matter causing an event or state of affairs in another) and by laws of nature (that is, there being a law of nature relating events or states of affairs in two or more subject-matters). But, say these authors, these subject-matters are not all somehow dependent on some basic subject-matter (that of physics), nor somehow arranged in a hierarchy or spectrum of 'basicness'.

My own position (of course, shared with many authors) is that the mind–matter distinction *can* be made good; and, furthermore, once it is made good, this last claim is wrong. That is, physics is indeed 'basic', and modern science is well unified. At first sight, this position is likely to seem 'reductionist' or 'eliminativist'. But as I develop this position in this section and the next, I will show that in fact it is not (at least in the most usual senses of these notoriously vague words!).

Given the distinction, there are, as I said above, various problems to address, some scientific and some philosophical. The most exciting scientific problem is perhaps (as suggested by Searle (in Chapter 2)) how mental states, especially conscious states, are biologically produced. I, of course, agree with Searle that, for this problem, the philosopher's role is primarily to help clear away some confusions. Thus also Locke, who said in the 'Epistle to the reader', in his *Essay* of 1690 that he was content to be 'an underlabourer ... clearing ground a little, and removing some of the rubbish that lies in the way of knowledge' (1972, p. xxxv).

As to the philosophical problems, I shall concentrate on what I take to be the main metaphysical issue: in what sense, if any, do neural states 'underly' mental states? Lipton (Chapter 11) proposes that the two main candidates for this relation of 'underlying' are: causation and identity (see also Boden, Chapter 1). In section 3, I will advocate the second candidate: identity. (Lipton summarizes some difficulties confronting this view, and recommends scientists to be content with causation. But, as will emerge, I agree with most of what Lipton says.) As for the mind–matter relation in general, however, I should first address a sceptical view: that in no sense do neural states underly mental states.

2.2   Denying that neural states underly mental states

This sceptical view is based on the point that mental states, as we ordinarily conceive them, involve the world: in the jargon, they are 'wide'. This width is shown by some standard examples from recent philosophy of mind. Thus authors such as Burge, Kripke, and Putnam argue that you could not believe that London is pretty, that water is wet, that arthritis is painful, just in virtue of your brain state. For you can only be credited with the concepts occurring in these beliefs—concepts such as London, water, and arthritis—if your environment has certain features, and if you are appropriately related to those features. In other words, a Doppelganger of you, living on another planet, would not have these concepts, no matter how much their neural states matched your neural states, unless certain features of their environment, and their relation to those features, similarly matched yours. Furthermore, these features of the environment need not be physical features, nor features that can uncontentiously be spelt out in physical (or, more generally, natural scientific) terms. They can be social features, such as the meaning of 'arthritis' in a language. And the appropriate relation need not be a physical relation; nor need it be a relation that can uncontentiously be spelt out in physical, or natural scientific, terms. It might be, for example, membership of a linguistic community (see Burge 1979; Kripke 1979; Putnam 1975).

I take it that these points are established by these authors' work. Thereafter, matters become difficult: it is very hard to say, for a given concept, exactly what is required of the environment, and of your relation to it, if you are to have the concept. For example, to have the concept of London, must you have had causal contact (however exactly that is defined!) with London? But, of course, these difficulties do not need to be solved in order to attack the idea that neural states underlie mental states. For that, the established points are enough.

Indeed, these very difficulties enable one to generalize the attack. Thus, it can seem plausible that however exactly you define 'mental' and 'physical', mental and physical facts are intertwined in the world in so myriadly complex a way, that no limited class of physical facts 'underlies' (in any sense) all the mental facts—and that this remains so, even when 'physical' has the very general meaning 'natural scientific', and even when one goes beyond the body of the individual person or animal. (In going beyond the individual, one immediately meets the social dimension of mind; here, talk of mental, and indeed social and cultural, 'facts' is more common than talk of 'states': hence this paragraph's change of terminology.)

In other words, it seems that not only do the mental facts about a person, or animal, 'outstrip' the facts about its nervous system: they outstrip all the physical, chemical, and biological facts about it—and even those of other persons or animals. To adapt Midgley's example of treaties (Chapter 8), it seems that no collection of natural scientific states of a nation's citizens underlie the nation's agreeing to a peace treaty.

This view is, of course, close to the sceptical view I discussed in subsection 1.1, that either the mind–matter distinction cannot be made or, once it is made, it is without significance. The difference is that this view is less radical. It allows that the mind–matter distinction may be made (perhaps in various ways), and have various kinds of significance. But it denies one much-discussed kind of significance—that the physical (even in the sense of 'natural scientific') 'underlies' the mental.

My reply to this view (again, shared with many authors) consists of two parts. Though I endorse both parts, they are independent: one can endorse each part without the other. The first part is directed at the generalized attack, and is, I admit, wholly programmatic. In effect, it consists of two broad strategies for finding the physical facts underlying some mental (or any other, for example, social) fact, such as acceptance of a peace treaty; as follows.

1. To find these physical facts, we may need to look far more widely across the environment, both in time and space, than is initially suggested by a sentence reporting the fact. Thus the physical facts underlying the social fact 'Napoleonic France accepted the treaty of the Congress of Vienna' will surely not be confined to France and Vienna, even in the year 1815. As to time, the physical facts underlying France (or any nation) accepting a treaty may well include facts earlier even than the lifetimes of the relevant politicians, or other citizen—who knows how far back in time one has to go, in order for the physical facts to fix our concept of a treaty? As to space, a similar point can be made.[2]

2. To find these physical facts, break up the problem. That is: think of some sequence of kinds of fact, which has some kind of physical facts as the first member of the sequence, and for which it is plausible to claim that each kind of fact underlies the kind that next occurs in the sequence—and then argue *seriatim* for these claims. As thus stated, this strategy suggests the traditional reductionist hierarchy, with physics reducing chemistry, chemistry reducing biology, biology reducing psychology, and psychology reducing the social sciences (Oppenheim and Putnam 1958; Monod 1972). But this strategy is by no means committed to this hierarchy, regarded by many as a bugbear, for at least three reasons (and others will emerge below).

First, 'underly' need not mean 'reduce' (at least in the sense of 'reduce' that you consider a bugbear!). Second, the strategy allows different sequences of kinds of fact (maybe criss-crossing), rather than just a single line going 'up' from the physical. Third, the strategy allows very different kinds of fact (as well as sequences) than this traditional list.

Here, it is worth mentioning a well-worked-out example of the strategy that uses such different kinds of fact (although the example is well known to analytic philosophers, it is almost unknown outside philosophy); namely, David Lewis's strategy for establishing what he calls 'Humean supervenience' (Lewis 1986a, pp. ix–xiv). Roughly speaking, his plan of battle is as follows. He argues that the physical facts underlie (a) facts about laws of nature and (b) facts about counterfactual conditionals, though neither of (a) and (b) underlies the other. Facts about counterfactual conditionals then underlie (c) facts about causation. Then (a) and (c) underlie (d) largely individualistic facts about the mind (via a version of functionalism). Then (d) underlies linguistic facts (here the argument includes an analysis of linguistic convention in terms of mental facts about, for example, common knowledge of how symbols would be interpreted). The arguments for these claims are impressively detailed—so much so that one can indeed believe that physical facts underlie a nation's accepting a peace treaty: I recommend them to Midgley and other sceptics. But we must forego the details of these arguments. For now, it is enough that we can already see how different are these kinds of fact from the traditional list above. (And the further details of Lewis's battle-plan reveal further differences.) To sum up: this strategy can be very imaginative!

I turn to the second part of my reply to the sceptical view (again, shared with many others). It is directed at the arguments and examples of Burge *et al.*, and is more specific. In contrast to the 'look more widely' of (1) above, it says in effect 'look more narrowly'. That is, I concede that mental states as ordinarily conceived are wide (as the examples suggest), but still claim that a person (or other animal) has *some* mental states just in virtue of their neural states. Such mental states—ones that would be shared with an organism with exactly matching neural states—are called 'narrow'. The usual candidates are states of sensory experience: something like seeing yellow in the top left of one's visual field. Agreed, such states are hardly mentioned in everyday life, but there is an obvious pragmatic reason for this. We are primarily interested in describing, deciding, predicting, and explaining our own and other people's actions (and here 'action' means not just the 'narrow' idea of body-movements, but these movements' causal consequences in the world). With actions thus widely conceived, describing, deciding, predicting, and explaining them naturally mentions wide mental states. Thus the advocate of narrow mental states can, and no doubt should, allow that such states are hard to describe in everyday language, which tends to use a 'wide' taxonomy of states. Maybe even 'yellow', as a word of everyday language, expresses a concept whose possession requires something of one's environment (no doubt, nothing so strong as having seen a lemon or banana—but maybe one has to have seen a yellow object).

But despite these difficulties of description, I think that there are such narrow mental states, and that states of sensory experience are examples; so that some narrow states form a constant accompaniment to the wide states that we usually ascribe to ourselves and other people.[3]

2.3 Physicalism with a human face

Given such narrow mental states, the way is clear to discussing the main metaphysical issue introduced above: exactly what is this relation of 'underlying', between neural states and narrow mental states?

In section 3, I will address this question: namely by formulating physicalism as a thesis of supervenience. Roughly speaking, physicalism will say that physics is indeed the 'basic' science. A bit more exactly: physicalism will say that as a matter of contingent fact the physical facts about the cosmos completely fix all the facts. This will make the relation of 'underlying' be identity, that is neural states are narrow mental states. (I will also suggest that given their other views, Searle and Lipton can and should agree.)

Before embarking on the details of formulating physicalism, however, I want to emphasize that my physicalism does not imply various dubious doctrines. First, it will turn out that my physicalism is not 'reductionist' or 'eliminativist', in most senses of these words. We can already state one reason why not. It lies in the distinction between metaphysics, on the one hand, and epistemology and explanation on the other. Thus my metaphysical thesis of physicalism, and the thesis that neural states are (narrow) mental states, implies very little about the epistemology of mental states (for example, how can we know about them, and what are our best methods for getting such knowledge?), and very little about the character of explanations of mental states and of processes (that is, sequences of states). In particular, it does not imply that physical explanations of these states or processes are somehow 'better'. A principal reason for this is the fact that, as Lipton rightly emphasizes (in Chapter 11), most explanations are contrastive; and once we recognize this, we can overcome the illusion that there is, or should be, a 'complete explanation' of something, such as a mental state (as he notes, this point of view is also endorsed by Boden, Midgley, and Watts—see Chapters 1, 8, and 9).

Similarly, I maintain that my physicalism and the identity of narrow mental states with neural states do not imply the doctrines that Rose identifies as the main errors of reductionism—reification, arbitrary agglomeration, and such like—and which he rightly castigates as not merely false, but as leading to grotesquely wrong public policies (see Chapter 5). I shall not tackle these errors point by point. Suffice it to say here that I think all but two of these errors arise from the naive desire to explain myriadly different phenomena in a simple, unified way, in terms of their smallest parts—a unitary model of explanation that Lipton shows us to be a chimera. The two exceptions are the errors of improper quantification and belief in statistical 'normality'. These obviously arise from our scientific culture's over-valuing mathematical models, which is itself caused, at least in part, by the success of mathematics within physics. But for all its grotesquely wrong social consequences, this over-valuing of mathematics is, from the strict viewpoint of the philosophy of the mind–matter relation, a 'merely' historical fact. And right or wrong, setting a high value on mathematical models is a methodological claim, and no part of the metaphysical thesis of physicalism.

3 Physicalism

I turn to formulating physicalism. It says, roughly speaking, that all empirical subject-matters—such as the biological, the mental, and the social—are literally a part of the subject-matter of physics. I shall make this precise as a so-called 'supervenience' thesis, where supervenience is a relation between subject-matters. Although several authors (including physicalists) have recently criticized this approach to formulating physicalism, I believe it is viable—and that physicalism so defined is plausible. But I have space only to spell out the approach (in a version that owes much to the metaphysical system of David Lewis), not to address the criticisms.

3.1 The material and the physical

Before explaining the crucial relation of supervenience (subsection 3.2 et seq.), I need to make three preliminary remarks, which will guide the precise formulation of 'physical' and of 'supervenes'.

1. A claim more modest than physicalism is often discussed: it is usually called 'materialism'. Again speaking roughly, materialism claims that all empirical subject-matters are part of the subject-matter of the natural sciences. So materialism does not require that they all be part of physics: one could be part of physics, another part of biology, and so on. Although this is weaker than physicalism, it is still quite radical. For, taking just the case of interest to us, the mind: this seems a very different subject-matter from those of the natural sciences. The two general differences most commonly cited in philosophy are that two concepts (properties) that seem important for describing minds and experiences seem entirely absent from all the natural scientific theories (in physics, chemistry, biology) that we use to describe the material world. Namely, the concepts of (a) intentionality—the concept that a belief or perception or desire or some other mental state is about an object or state of affairs; and (b) qualia—the concept that a conscious mental state, like a perception of an object outside the body, or a bodily sensation, has a 'raw feel', 'phenomenal quality', or 'quale'. Materialism (and *a fortiori*, physicalism) is committed to claiming that these differences are illusory: that a natural scientific account of intentionality and qualia is possible. To put the same claim in terms of subject-matter, instead of its linguistic description: the concepts (properties) that seem distinctively mental are in fact material—though perhaps very complicated, or in some other way special.

2. Materialism has been widespread in philosophy since the late nineteenth century, as a result of the great success, since about 1800, of natural scientific theories in predicting and explaining the material world. One well-known

example is the demise of vitalism within biology. Here are some examples from astronomy, chosen with a view to emphasizing how it gradually emerged that the same laws govern processes on Earth and far away in space—a remarkable unity in the material world, which nowadays we tend to take for granted. In 1793, Herschel showed that double stars circle each other, confirming that Newton's law of gravitation applies outside the Solar System. From 1860, Kirchoff and Bunsen applied their spectroscopes to sunlight, and thus paved the way to determining the chemical constitution of the stars (just 25 years after the positivist Comte gave this as an example of the sort of information that science could never attain!). And it emerged that all the elements detected in the stars are also present on Earth; though, to be sure, there were times when this seemed false, the best-known example being the 27-year gap between the discovery of helium in the Sun (1868: hence the name), and on the Earth (1895).

Similarly, physicalism is widespread in present-day philosophy, as a result of the great success, since about 1900, of physical theories in predicting and explaining not only physical, but also chemical and biological processes. Here are two well-known examples, chosen with a view to emphasizing how quantum theory can now claim to be the fundamental theory of matter— and how this claim has been hard-won: first, quantum theory's explanation of the homopolar chemical bond (achieved in 1927, just after the discovery of quantum theory, by Heitler and London); second, quantum theory's explanation of superfluidity and superconductivity (achieved by 1960, by the combined work of many).

Of course, I do not intend these historical points as persuasive arguments for the truth of materialism and physicalism. But I do think they indicate that precise formulations of materialism and physicalism should render them as contingent theses. For it is a thoroughly contingent fact that since 1800 there has been such supreme success in the natural scientific description of the world; and that since 1900 there has been such supreme success in physics. So surely our formulations should reflect this contingency of our intellectual cultures.[4]

3. This leads to the next remark. Today's natural scientific theories, and physical theories, are, of course, not wholly true. No doubt some of what they say is false. Furthermore, they are partial: they do not completely describe (truly or falsely) their own immediate subject-matter, let alone subject-matters like the mental and the social that materialism and physicalism claim to fall within their scope. So materialism and physicalism should be formulated so as to claim correctness and completeness only for some sort of corrected and extended versions of these theories. So the question arises:

how exactly should we define these corrected and extended versions? The threat of triviality looms: we must not define them as whatever corrections and extensions are needed to make our formulations come out true (as opponents of physicalism have often pointed out: for example, Healey 1978; Crane and Mellor 1990).

This is an important, and hard question, which I cannot answer exactly. It must suffice to make a few points. First, some authors sketch definitions in terms reminiscent of Peirce: they appeal to the long-run future of the natural sciences, or of physics—or, rather, what this future would be, if circumstances were sufficiently propitious for enquiry (such as enough funding). But I find such definitions too dependent on contingencies about what humans require for successful enquiry. I prefer more metaphysical definitions: specifically, those based on the claims that (a) we already have, from present-day theories, a good idea of what counts as a natural scientific, or physical, property and (b) we can readily enough make this idea precise. Given claims (a) and (b), the corrections and extensions can be straightforwardly defined as the true complete theories of those properties.

So the issue turns on the claims. Obviously, the greater unity of physics, as against the various natural sciences taken together, makes claim (a) more plausible for physical properties, than natural scientific ones. (Because I favour physicalism, this difference of plausibility is no difficulty for me: materialism, being logically weaker than physicalism, is supported by whatever evidence supports physicalism.) I myself believe that our present idea of physical property has two main components; I admit that both components are vague (the second more so), so that establishing claim (b) is hard—but for the present, they will have to do.

First, a physical property is a numerically measurable property of objects, whose value changes in time (if at all) according to a law that relates it to other such properties and their values (typically, a differential equation or generalization of such, like a stochastic differential equation). Here, and in what follows, 'properties' includes relations such as 'rotating faster than' as well as monadic properties such as 'is electrically neutral'; and 'objects' is meant very generally, including what we would more naturally call events, processes, or states of affairs.

Second, any such property is related to the physical properties we have already discovered, the number of which is astonishingly small, given the universal scope of physics: depending on how you count, there are somewhere between about a dozen (including position, mass, electric charge, and spin) and a hundred (including energy, momentum, entropy, temperature, and conductivity). This relation to the already discovered properties is to be (or include) numerical relations between values, reported in equations. (This relation might be reduction or supervenience as defined below, but it need not be: physicalism should, of course, allow that a genuinely new physical property—one not supervening on the already discovered—might exist.)

3.2   Relations between subject-matters: reduction and supervenience

So much for preliminary remarks. I turn to being more precise about relations between subject-matters. In accordance with remark (3) in subsection 3.1 above, I take a subject-matter to be a characteristic family (a set) of properties, defined on some set of objects. (Again, 'properties' includes relations; and 'objects' includes, for example, events and states of affairs.)

You might reasonably object that a subject-matter should include not only a set of objects, and a family of properties, representing a taxonomy (classification system) for the objects but also doctrines—general propositions about how the properties are related; for example, that all Fs are G, that no G is both an H and a K (such propositions might even be laws). But no worries: I shall take the subject-matter to include each property, not just as a 'concept', but rather as an extension (a set of instances). Since all such general propositions will be implicitly fixed by the extensions—so, for example, the set of Fs is a subset of the set of Gs—they will in effect be included in the subject-matter.

Similarly, it will be convenient to speak below of two properties being identical with another, when they are co-extensive (have the same set of instances) in the given set of objects. At first sight, this usage seems to violate the standard view that intuitively distinct properties can be accidentally co-extensive; for example, 'has a kidney' and 'has a heart'. But it is just a usage: I endorse this standard view. Furthermore, I will argue in subsection 3.4 that for the case of interest (physicalism) the set of objects needs to go beyond the actual world—to include objects that do not actually exist. As a result, this usage will not even appear to violate the standard view.

As mentioned in subsection 2.1, one obvious way in which two subject-matters can be related is by causation and/or law, while neither is in any sense part of the other. Thus there might be causal relations between objects in the two sets; and there might be non-causal but nomic relations (from the Greek word for law, $v\omega\mu\omega\sigma$). An example is classical electricity and magnetism; where each is thought of as, say, the family of possible values for the electric (or magnetic) field, defined on the set of spacetime points. There are certainly nomic relations between these values (given by Maxwell's equations); and maybe causal ones too, though the role of causation in physics, even classical physics, is controversial. But neither subject-matter is a part of the other (though indeed, both are reduced to a single underlying subject-matter, the electromagnetic field). An example from everyday experience is France and

England; where each is thought of as, say, the family of all empirical properties of the relevant set of citizens. Here there are plainly causal, as well as nomic, relations; but neither is part of the other.

For one subject-matter to be part of another, it is obviously necessary that the first subject-matter's set of objects be a subset of the second's. But that is not enough: in the example of electricity and magnetism, there is a single set of objects—the spacetime points—but two distinct subject-matters. Obviously, the first subject-matter's family of properties must also be somehow a part of the second's: every classification of objects made by the first must also be made by the second. The simplest way this can happen is if the first's family of properties is included in (that is a subset of) the second's. (Since I take a subject-matter to include each of its properties' extensions, this inclusion will be enough to secure that the doctrines—general propositions—of the first subject-matter are included in those of the second.)

But, typically, we are not so lucky: subject matters are often given to us without one family being a subset of the other. A trivial example is squares and rectangles. Because squares are just a special case of rectangles, the subject-matter, squares, should surely be part of the subject-matter, rectangles. But the latter might not be given to us as including the property 'is a square', and all the various related properties, like 'is a diagonal of a square', that occur in the subject-matter squares. After all, why should the subject-matter, rectangles, single out such special cases?

The remedy is not far to seek. We obviously need to use the idea of *compounding* properties to yield other properties, where the compounding operations can in general be iterated. Then we define one subject-matter to be a part of another if and only if:

(1) its set of objects is a subset of the other's; and

(2) its properties either are among the other's properties (the simple, lucky case) or are compounds of them.

(As I said above, clause 2's notion of identity for properties and their compounds is straightforward: it is just coextension in the given set of objects.)

This definition focuses attention on compounding operations. About these, I should first make two brief, related points. They both arise from my taking a subject-matter to include each of its properties' extensions. First, it turns out that for the usual compounding operations, clause (2) of the definition implies clause (1). Second, more important, we should no doubt require that if one subject-matter is a part of another, then its doctrines are in some corresponding sense a part of that other's; for example, by being entailed by them. And you might reasonably worry that, even with subject-matters taken as including extensions, the above definition does not secure this requirement—might not compounding break out of the class of entailed propositions? But it turns out that for the usual compounding operations, there is no problem: the definition implies that the doctrines of the first subject-matter are indeed entailed by those of the second.

So what are these compounding operations? The most simple are the so-called Boolean operations, represented by words such as 'and', 'not', and 'or', giving, when iterated, all the Boolean compounds of the initially given properties. These operations will certainly suffice in simple cases like squares and rectangles, maybe even without iteration. For example, suppose the subject-matter, rectangles, is given as containing the properties, being a rectangle, and being a plane figure with four equal sides. The conjunction of these properties is the property, being a square. And once this link is made among properties, there is, of course, no problem about the deduction of the doctrines: you can deduce all doctrines about squares from all the doctrines about rectangles.

But there are more complex compounding operations. The two obvious examples from logic and philosophy are applying the quantifiers 'all' or 'some' to get a monadic property, such as 'bears relation R to everything/something', from the binary relation 'bears relation R to ' (or, more generally, an n-adic relation from a $(n+1)$-adic relation). These two examples behave rather like the operations of conjunction and disjunction ('and' and 'or'), respectively. But they cannot be finitely defined in terms of conjunction and disjunction, because they make sense even if there are infinitely many objects. In that case, 'all' and 'some' behave like infinitely long conjunction and disjunction.

This analogy might make the quantifiers seem a small addition to the Boolean operations. But they are significant in two ways. First, there are striking examples of a subject-matter being rigorously shown to be part of another, in the above sense—once we allow quantifiers; with the attendant deduction of its doctrines from those of the other. The outstanding example is the demonstration (usually put in terms of deducing doctrines, rather than compounding properties) that all of classical pure mathematics is part of the theory of sets. This was a remarkable achievement, taking half a century of effort (from about 1860 to 1910) by many mathematicians.

Second, infinitely long conjunction and disjunction suggest the more general idea of iterating any operation, infinitely rather than finitely. This is significant for us. Take any well-defined collection of operations for compounding properties. (There are indeed others, not definable just in terms of the Boolean operations and quantifiers; for example, interpolation in a spectrum of properties or extension of such a spectrum.) Then there are clearly two main ways to interpret the above definition of one subject-matter being part of another; as follows.

Either one restricts oneself to finite iterations: this I call 'reduction' though, as noted above, this term has many other (and vaguer) uses. This is, of course, the way you read the above definition: and it applies to the two examples above—squares and rectangles on one hand, and pure mathematics and set theory on the other. Or one allows infinite, as well as finite, iterations. That is mind-stretching and requires care, if paradoxes are to be avoided: but it turns out to be tractable. We say 'infinitary' for 'infinite or finite'; so that this interpretation of the definition is sometimes called 'infinitary reduction' (just as the corresponding branch of logic is called 'infinitary logic'). It is important because it turns out to be equivalent to a notion that is much discussed—so called 'supervenience'.

The exact definitions of 'supervenience' vary from one author to another. But the common, main idea is that one family of properties supervenes on another, with respect to a given set of objects (on which both families are defined), if any two objects that match for all properties in the second family (both having or both lacking each such property) also match for all properties in the first family. Or, equivalently, the contrapositive formulation: any two objects that differ in a property in the first family (one object having the property, the other lacking it) must also differ in some property or other in the second family. It is straightforward to show that these definitions are equivalent to infinitary reduction, as defined above.

A standard, largely uncontroversial example of supervenience is given by paintings. Most agree that the family of aesthetic properties of paintings (properties such as 'is well composed' and 'has the colouring of a Matisse', or even the hack example, 'is beautiful') supervenes on their pictorial properties; where by 'pictorial properties' I mean some set of non-evaluative properties of tiny regions, such as 'is magenta for the one square millimetre in top left corner'. Of course, if supervenience is to hold good, the family of pictorial properties must be suitably rich. In particular, they cannot just describe colour, even in a technical vocabulary, millimetre by millimetre; they must also describe details of the medium, such as oil or watercolour. But most would say that a family of non-evaluative properties can be picked out such that if two paintings matched utterly in respect of these properties, then (no matter how else they differ—maybe one is the original and the other is a fake), they match in their aesthetic properties: if one is well composed, so is the other, and so on.

This example also illustrates the contrast with (finite) reduction. Is a property such as 'is well composed' a finitely long compound, built out of the pictorial properties? Many who are happy to accept supervenience have thought not: they point out that we hardly know how to begin writing such a finite definition, let alone how to improve it and perfect it. But here a warning is in order. Because a finite definition of 'is well composed' could be so long (for example, a million million pages) as to be incomprehensible by human minds, even working collaboratively, our being unable to begin writing such a definition is very weak evidence for its non-existence. So reduction might yet hold.

3.3   Formulating physicalism

We can now move towards formulating physicalism. For clarity, I shall first state the main idea, and then make it precise. The main idea will, of course, be that the mental reduces to, or at least supervenes on, the physical, in the above sense: that is, for reduction, each mental property is coextensive with some physical property, one of the simple given ones or a finite compound. For example, imagine that there is a compound (finite but perhaps very complex) physical property X such that any person or animal is an instance of 'sees yellow' if and only if they are an instance of X. Then 'sees yellow' is reduced.

The main idea of supervenience, then, is: for any two objects, if they match on all their physical properties (for any physical property, they both have the property or they both lack it), then they match on all their mental properties. For example, consider a person, who, during 2 seconds, sees yellow. Imagine a physical duplicate (replica) of that person: this means that corresponding physical parts of the person and of the duplicate, right down to the smallest parts (the level of atoms or electrons or smaller still), are to be in identical physical states. And imagine that we control the environment of the person and of the duplicate, so that the duplicate remains a duplicate, moment by moment, for 2 seconds. (No doubt, to do this we will have to make the duplicate's immediate environment be itself a duplicate of the original person's immediate environment—a hard assignment.) Then supervenience implies: the duplicate also sees yellow.

So far as I can tell, most people, when asked whether such a duplicate would see yellow, confidently answer 'yes'. I confess: being a philosopher rather than an empirical sociologist, my evidence is anecdotal—when I say 'most people', I mean, strictly speaking, 'most philosophy undergraduates whom I have asked in a classroom survey'. But I ask them this question early in the course, before I make any attempt to persuade them that physicalism is true. So, though anecdotal, this evidence supports the point at the beginning of this section: that physicalism is widespread in our intellectual culture. On the other hand, these students' opinions are more divided, and, of course, more tentative, about whether reduction holds—that is, whether there is a finite definition of 'sees yellow' in terms of physical properties. But the above warning is again in order. Because a finite definition of 'sees yellow' could be so long as to be incomprehensible, our inability to write such a definition is weak evidence for its non-existence: reduction might yet hold.

So much for the main idea. To formulate physicalism precisely, we need to decide on two points: exactly what is the set of physical objects, of which the sets for other subject-matters will be a subset (compare clause (1) of the definition in subsection 3.2); and exactly what compounding operations among properties to allow (clause (2)). Almost all of the literature focuses on the first point, though the second is obviously just as important; but I shall follow the literature, as I have nothing useful to say about the second point.

At first sight, it seems that clause (1) will require only that all the objects on which mental properties are defined are objects on which physical properties are also defined: that is, the mental subject-matter's set of objects is to be a subset of the physical subject-matter's set. And this seems uncontroversial. For we normally think of mental properties as defined on people and animals; for example, we say 'Fido sees yellow'; and people and animals also have physical properties.

But this clause is not so straightforward, for three reasons. The first two are closely related to the discussion in subsection 2.2; the third is unrelated to previous discussion, and will lead us to the next subsection.

1. Two subject-matters, say the physical and the mental, are typically presented to us with properties that do not have common instances. For example, '... sees yellow' has Fido as an instance, even when we interpret '... sees yellow' as expressing a narrow mental state (compare section 2.2). But the physical properties that a physicalist claims to 'underly' this narrow mental state are typically presented as properties, not of dogs but of brains or parts of brains: the standard example, in the philosophers' pretend-technical jargon, is '... is a firing of C-fibres'.

So to satisfy clause (1) in subsection 3.2, we must expect to have to alter somewhat the properties (and so also the predicates expressing them) that are presented to us. In the example, either we must introduce a mental property, corresponding to '... sees yellow', defined on brains or their parts, as it might be '... is a part of a brain in the narrow state of seeing yellow' or we must introduce a physical property, corresponding to '... is a firing of C-fibres', defined on organisms, as it might be '... has a part of its brain that is firing its C-fibres'; or we can, of course, do both. To sum up: to satisfy clause (1) we must expect to do two things: make our set of objects contain parts of its own members, and/or wholes composed of its members; and 'massage' the properties accordingly.

2. I said at the start of this section that physicalism claims all subject-matters, not just the mental, to be part of the physical. So clause (1) will

require that all the objects on which any properties (such as biological or social properties) are defined are objects on which physical properties are also defined—that is, are members of the physical subject-matter's set of objects. Indeed, even if one took physicalism to claim only that the mental is part of the physical, the social nature of mind, emphasized in section 2, would prompt physicalism to put social objects, such as linguistic communities and nations, into the physical subject-matter's set.

This, of course, leads us back to the first reason, just above. The physicalist must somehow conceive objects such as linguistic communities as physical (quite apart from issues about their properties). The obvious strategy for the physicalist is as just above: first, assume set-theory (or a similar device such as mereology, the theory of parts and wholes); second, conceive these objects as sets (or mereological wholes) with smaller, more obviously physical, objects—such as organisms or limbs or organs or cells or molecules—as members (or parts); third, take the set of physical objects as containing all sets (or wholes) composed out of objects that are given as being physical.

3. The third reason for care about clause (1) leads us to the topic of modality (the notions of possibility and necessity). Here I shall show why we have to deal with these notions. In the next subsection, I will adopt a widespread, though admittedly controversial way of doing so.

Clause (1) says 'a set of objects'. One naturally reads that as a set of actually existing objects, with their actual properties. But if we only consider the actual world (the actual total history of the Universe, throughout all time and space), there is a grave risk that reduction—and even more risk that supervenience—will be trivially true; and true for a reason that has nothing to do with the intuitive idea of physicalism. So there is a risk that our formulations will not capture this intuitive idea.

The obstacle is very simple. Here is an example for reduction of a mental property. Suppose that, as it happens, the set of actual instances of 'sees yellow' is finite: that is, throughout the actual world (in the above sense), only finitely many people and animals see yellow. Then there surely is a compound physical property with exactly that set as its instances. To find such a property, I could find for each instance of 'sees yellow', a physical property possessed only by that instance; and then form the disjunction of those properties. This finite disjunction will have exactly the yellow-see-ers as instances. Then 'sees yellow' is reduced. (For simplicity, I have stated the problem in terms of whole organisms as instances, and regardless of the issue of change—seeing yellow at one time but not another. The problem is unaffected by these points.)

Here is an example for supervenience. Suppose that no two actual people or animals exactly match in their physical properties (that seems extremely likely). Then supervenience is trivially true, on the usual understanding of 'all' and 'if, then'; as logicians put it, it is 'vacuously true'. (Again, the problem is unaffected by my choosing organisms as instances, and by my ignoring change.)

But the truth of physicalism—whether it be reduction or supervenience —must not be established by the mere accident of the reduced property having finitely many instances, or of no two objects perfectly matching in the 'subvening' properties. So something has gone wrong with our formulation. (Being true on account of such a mere accident is, of course, sufficient for being contingent, which I urged in subsection 3.1 was a desideratum in formulating physicalism. But this desideratum should not be earned so cheaply: the accident of finitude or lack of match is clearly irrelevant to the intuitive idea of physicalism.)

3.4 The range of supervenience: going beyond the actual world

We need a formulation of physicalism that somehow circumvents the mere accidents seen at the end of subsection 3.3. That must mean a formulation that is actually true, according to the physicalist, on account of how mental and other properties would be reduced, even if there were infinitely many instances of them (or would supervene, even if there were perfect matches in physical properties). Such talk of how things would be, in contrary-to-fact circumstances, is often treated in terms of possible worlds. I will follow this tradition.

We can introduce possible worlds as follows. Many true propositions are contingently, not necessarily, true. The actual world makes the proposition true; but, as we say, 'it did not have to be true'. Following Leibniz, and modern modal logic, we say: there is a possible world (different from the actual world) at which the proposition is false. So we imagine the set of all logically possible worlds. The actual world is one of them. They all make true all necessarily true propositions: the propositions of logic, maths, and analytic propositions such as 'All bachelors are unmarried'. And so also, in general: for any proposition there is a set of possible worlds at which it is true. (Any contradiction, or more generally impossible proposition, is associated with the empty set of worlds.) And any two logically equivalent propositions are associated with the same set of worlds. (For details of possible worlds' uses in philosophy, and debate about exactly what they are, see Kripke 1980; Lewis 1986b).

Given this framework of possible worlds, reduction or supervenience will take as the set of objects a set of actual and possible objects. The absorption of other subject-matters by the physical, that reduction and supervenience claim, will be logically stronger, because holding on a larger set. For example, even if no two actual people or animals exactly match in their physical properties, surely there could be a person who exactly matches my present physical properties. So there is a possible world containing such a person. So if supervenience takes such a person in such a world, together with me, as members of its set of objects, then supervenience requires that I and that person match in our mental and other properties.

Here I have again taken an example with organisms as the objects (instances of the property) in question. I will continue to concentrate on this case. But, as above, my discussion will be unaffected by this choice. We could accommodate the 'width' of the mental and social by having large regions of spacetime—or even entire possible worlds—as the objects. With these alternatives, authors often speak of 'regional' and 'global', supervenience, respectively; see Kim (1984, p. 168) and Horgan (1982, pp. 36–7).

So the question is: exactly which set should be taken as the set of objects? Presumably, both reduction and supervenience should take, for any given possible world, either all the people and animals in that world, or none of them. (And similarly for organisms' parts, and wholes composed of many organisms; see subsection 3.3.) So, which worlds should be considered?

This is a hard question, whose resolution depends on deep and controversial issues unrelated to mind: for example, issues about laws of nature, and the identity-conditions for properties. I can do no more than broach the issues. I shall do this by discussing two simple choices of worlds: choices that relate to the discussion above and to the quantum theory below.

1. Choose the set of all logically possible worlds. With this choice, reduction would be a matter of necessary coextension of properties. Supervenience would be a matter of: any two logically possible items that physically match also match mentally and in all other respects.

I reject choice 1 because it makes physicalism either logically necessary or logically impossible (for it makes physicalism a matter of a pattern across all the worlds, and thereby true at all worlds or at none). Moreover, I said in subsection 3.1 that I believed physicalism should be formulated so as to be contingent. But I should remark that some authors are content with choice 1, and so endorse physicalism as necessary.[5]

There is another general advantage in rejecting choice 1. It concerns the identity of properties: namely, rejecting choice 1 gives a physicalist, even a believer in reduction, ample scope to agree with non-physicalists that mental, social, and other properties are *not* identical with physical ones. The reason is that most philosophers agree that for two properties to be identical, they must be necessarily coextensive. At least, almost all philosophers who are willing to talk at all about identity-conditions for properties agree on this (they then dispute whether necessary coextension is also a sufficient condition for identity of properties). If we agree with this, then a physicalist who rejects choice 1 is accepting that mental and other properties are not identical with physical ones; for a necessary condition of such identity is violated.

This advantage of rejecting choice 1 is uncontentious. But it is worth emphasizing, for two reasons. First, it may sugar the pill of physicalism for some who find it incredible that a mental property could be a physical property (Lipton is such a person; see Chapter 11). The sugar is that they do not have to believe this: physicalism requires only a contingent coextension —albeit over some decently wide class of worlds (yet to be specified), so as to avoid the risk of merely accidental truth.[6] Second, this advantage of rejecting choice 1 may seem to conflict with my own talk about properties being identical when they are coextensive. But recall my remarks at the start of subsection 3.2: I warned that my usage was convenient but unusual—and was not intended to deny the standard view that contingently coextensive properties are not identical.

2. The second choice for the set of worlds invokes the idea of a law of nature. Many philosophers agree that the actual world has laws of nature; and that these are, or at least correspond to, contingently true universal generalizations. For us, nothing turns on the difference between 'are' and 'correspond to'. The main point is that laws differ from the merely accidentally true universal generalizations (for example, 'the coins in my pocket are silver') by being in some way very informative about the world; though the exact nature of this informativeness is hard to analyse, and is controversial.

Given this notion of law (however 'informative' is analysed), there is a well-defined set of all worlds that make true all the actual world's laws of nature. It is a subset of the logically possible worlds, because laws are contingent; a subset that includes the actual world—because the laws are actually true. This set is often called the (set of) nomically possible worlds (again, after the Greek word, $\nu\omega\mu\omega\sigma$, for 'law').

I shall take this set as the second choice that physicalism might make. So reduction would then be a matter of nomic coextension of properties. Supervenience would then be: any two nomically possible objects that physically match also match mentally (and in all other respects).

As is obvious from the discussion of choice 1 above, this choice has the merit of having physicalism avoid claims of property-identity. It also makes physicalism contingent, whether as reduction or supervenience. This arises from each law being a contingent proposition (and, indeed, on some theories, a law might be a true proposition at a given world and yet not be a law there). So different worlds can have different collections of laws; so there may well be logically possible worlds where the laws allow physical duplicates to differ mentally—say by having mental laws that are unrelated to the physical laws.

I believe that physicalism, once formulated as supervenience with this choice of worlds, is plausible. In particular, I believe it can accommodate the peculiar features of consciousness; pre-eminently, conscious states' intentionality (their representing objects and/or states of affairs) and their subjectivity (their involving 'phenomenal qualities', or 'qualia').[7]

But I cannot enter details. I have no space; and by and large, I have nothing to add to excellent discussions elsewhere. But let me single out Dennett's (1991, ch. 12) and Lewis's (1990) arguments that physicalism can accommodate qualia. Although Dennett does not defend a precise formulation of physicalism, let alone one involving possible worlds, Lewis does. He favours a definition of physicalism close to choice 2 above. The principal difference is that he uses his theory of natural properties; he also copes with the 'width' of the mental and social by taking entire worlds as the objects. (For both differences, compare (M5) Lewis (1983, p. 364), while choice 2 corresponds to his (M4).) Despite this difference, Lewis's argument about qualia can be easily adapted to choice 2: his central idea—that knowing what an experience is like is a kind of knowing how, not knowing that—is unaffected by the difference.

There is, however, a problem with choice 2. Although I have no space to discuss it is detail, I should describe it. For in subsection 5.2, I shall show how it is illustrated by an interpretation of quantum theory—giving my second conclusion, as announced in section 1. The problem is that a world where intuitively physicalism is false, because there are mental properties that are intuitively not supervenient nor reducible, may yet be a world where my formulation, with choice 2, is true—because the mental properties are correlated with physical properties according to strict (exceptionless) laws of nature. That is, choice 2 seems too weak; physicalism so formulated is too easily made true.

This problem is well known in philosophy (it is often noted in discussions of specific physicalist theories such as mind–brain identity theory and functionalism). There are two broad strategies for solving it. The first is conservative: keep the idea of formulating physicalism as a supervenience thesis, but make a different choice of worlds—for example, in the way proposed by Lewis (1983). The second is radical: give up supervenience and try to make sense, in some quite different way, of the idea that mental properties are 'constituted' by physical properties. In the literature, this second strategy is gaining ground. But there is much disagreement—and, I think, vagueness—about this new sense of 'constitution'. (Crane (1995, pp. 212–13) gives references for this strategy; he also argues that this strategy cannot accommodate mental causation.) Hence my preference for the first strategy. But I cannot justify this here: it must suffice to notice the problem, and the two strategies.

So much by way of formulating physicalism. Before turning to more details about physics, let me summarize the story so far by emphasizing the four main reasons why physicalism, as I have formulated it, is not 'reductionist' or 'eliminativist' about the mind (or, indeed, any other subject-matter).

1. Reduction and supervenience, in my senses, clearly legitimate, rather than eliminate, the reduced or supervening subject-matter—squares, and talk of them, are legitimated, not eliminated, by being shown to be part of the subject-matter, rectangles.

2. For a reduction in my sense to be useful for science, it must be manageably short; but this discussion, tilted as it is to metaphysics, has allowed a reduction to be arbitrarily long.

3. Many believe that for reductions (in any sense) to be eliminative and/or useful, the reducing theory (or subject-matter or what-not) must provide explanations of the phenomena that are treated by the reduced theory. Maybe so: but my physicalism is not committed to such physical explanations of mental phenomena; it is clearly compatible with pluralism about explanation (à la Lipton; see Chapter 11)—a pluralism that I in fact endorse.

4. Physicalism does not require the identity of mental with physical properties, at least in the usual sense that involves necessary coextension. It requires only coextension across a suitably large class of possible worlds, including the actual world.

4 Quantum theory and its measurement problem

But physicalism is only as precise as the word 'physical'! At the end of subsection 3.1, I discussed how to make this word, and especially 'physical property', precise in terms of similarity to the known physical properties. But however successful that strategy might be for formulating physicalism, it leaves untouched two dangers, which I will treat in two subsections. The first, lesser danger concerns why people believe physicalism is true. The second danger concerns its being true, and is our main business. It arises from quantum theory's measurement problem.

4.1 Naivety about physics

The first danger arises from a cluster of views, widespread in our intellectual culture, about physics as a science: that it has as its subject-matter, matter in motion, which it describes with precise mathematics; that what it tells us about this subject-matter is cumulative (that is it never gives up previously established claims); and even that at any given time, discussion among practitioners is uncontroversial. In short, the picture is of physics as a concrete floor of established, precise facts about simple concepts of matter and motion: a floor so firm (albeit perhaps dull) that other sciences can build upon it. Needless to say, this picture is false. Physics is much more interesting than this picture suggests. It has a much more varied, and strange, subject-matter than the matter in motion of classical mechanics; and it is steeped in controversy. (For an antidote to this false picture, see Leggett 1987, especially chs 5 and 6.)

But though false, this picture may well influence people (or at least my philosophy undergraduates) to think that physicalism is true, at least in the sense of supervenience. Here I have in mind a general, and a specific, point. The general point is that the usual verdict, in the thought-experiment about the physical duplicate of the person seeing yellow—that the duplicate also sees yellow—may well be influenced by the picture of people (and other organisms) or, more specifically, their brains, as composed of countless little particles.

The specific point is that a traditional argument against interactionism is flawed, because of this false picture of physics. Interactionism is the view, mentioned in subsection 2.1, that mind and matter are connected by causation and/or law, but neither is reduced to the other. The traditional argument against it has various forms; but it is often presented in terms of energy. The idea is that any causal interaction between mind and matter would violate the principle of the conservation of energy. Thus, if irreducible mental properties (or states, events or what-not) caused physical ones—for example, in wilfully raising my arm—that would surely mean that energy would flow into the realm of the physical. Similarly for physical properties or states causing irreducible mental ones: surely energy would flow out of the physical realm. But, says the argument, physics tells us that energy is conserved in the sense that the energy of an isolated system is constant, neither increasing nor decreasing. When it seems to change, the system is in fact not isolated, but rather gains energy from its environment, or loses energy to it. And there is no evidence of such energy gains or losses in brains. So much the worse, it seems, for interactionism. (Though traditional, the argument is still current; for example, Dennett endorses it (1991, pp. 34–5).

This argument is flawed, for two reasons. The first reason is obvious: who knows how small, or in some other way hard to measure, these energy gains or losses in brains might be? Agreed, this reason is weak: clearly, the onus is on the interactionist to argue that they could be small, and indeed are likely to be small. But the second reason is more interesting, and returns us to the danger of assuming that physics is cumulative. Namely, the principle of the conservation of energy is not sacrosanct. The principle was formulated only in the mid-nineteenth century; and although no violations have been established hitherto, it has been seriously questioned on several occasions. It was questioned twice at the inception of quantum theory (namely, the Bohr–Kramers–Slater theory, and the discovery of the neutrino). And, furthermore, it is not obeyed by a current relevant proposal (which I will discuss in subsection 5.1): a proposal for solving quantum theory's measurement problem.

In short, physicalists need to be wary of bad reasons to think physicalism is true, arising from naivety about physics.

4.2   Avoiding an indefinite macrorealm

I turn at last to quantum theory (from here on, QT). As I said in section 1, QT has an interpretative problem, called the measurement problem. There are many different strategies for solving this problem, but several are relevant to the mind–matter relation—more specifically, to physicalism. In the rest of this section I will sketch the measurement problem, and how it bears on the mind–matter relation. In section 5 I will discuss some strategies for solving it.

Roughly speaking, the measurement problem is this: QT's laws about how the states of objects change over time seem committed to the prediction that macroscopic objects often have no definite positions—nor definite values for other familiar physical quantities such as momentum or energy. ('Quantity' is jargon for 'numerically measurable property'; 'magnitude', 'variable', and 'coordinate' are also used.) But this seems manifestly false: tables and chairs surely have definite positions, and so on. As it is sometimes put: the macrorealm is definite. (Or at least, we experience the macrorealm as being definite. So, if QT is to account for our experience, it must either secure such definiteness, or at least explain the appearance of it. But I postpone till subsection 4.3 this second strategy; that is, the idea of allowing an indefinite macrorealm and securing only definite appearances.)

This problem is called the 'measurement problem', mainly because the argument that QT implies an indefinite macrorealm is clearest for a measurement situation. For QT says that microsystems, like electrons and atoms, in general do not have definite values for quantities. And if you use QT to analyse a measurement of, say, the momentum of an electron, which QT says has no definite momentum, you find that according to QT, the indefiniteness of the electron's momentum is transmitted to the apparatus's pointer—so that it has no definite position. I turn to spelling this out.

Like any physical theory, QT assigns states to systems: the state fully specifies the properties of the system. ('System' is just jargon for 'object'.) But the orthodox interpretation of a quantum state is as a catalogue of probabilistic dispositions. That is, for each quantity (position, energy, momentum, etc.), the state defines a probability distribution on all possible values of the quantity. These states are represented by vectors: they are often written (in Dirac's notation) with angle-brackets, e.g. $|\psi>$. For each state, there are some physical quantities and some value of each such quantity, such that: the state ascribes probability 1 to that value for that quantity. The state is an 'eigenstate' of the quantity, the value an 'eigenvalue'. But for each state, the great majority of quantities are ascribed a non-trivial probability distribution. This distribution is coded in the geometry of the vector space: the state is a vector sum of the quantity's eigenstates. And it is called a 'superposition' and is written with a '+'.

So far, so good. But the orthodox interpretation of QT adds that the system has a value for a given quantity *only when* its state ascribes probability 1 to that value. This is called the 'eigenvalue–eigenstate link'.[8] It is this scarcity of values that leads to the measurement problem. For the interaction of, say, an electron that is in a superposition (not an eigenstate) for momentum, with an apparatus for measuring momentum, leads to the electron's indefiniteness being transmitted to the apparatus—so that its pointer is in no definite position. This suggestion has been proven for a wide range of exact quantum theoretic models of measurement. But I will confine myself to a very simple model.

As an example, I will take a momentum measurement on an electron in a superposition of two momentum eigenstates: one for 1 unit of momentum, and the other for 2 units of momentum. Suppose we have a measurement apparatus or pointer, with 'ready state' $|r>$, which reliably reads these eigenstates, in the sense that the composite system behaves as follows:

$$|1>|r> \rightarrow 1>| \text{ reads '1'}> \qquad \text{and} \qquad |2r> \rightarrow |2>| \text{ reads '2'}>.$$

Here, the juxtaposition of two kets represents a state of a composite system, in our case the electron + pointer: namely the 'conjunction' of the two juxtaposed states. And the arrow represents the evolution of the state in time, as prescribed by QT's famous Schrödinger equation. So each of these displayed formulas means: if the composite electron + pointer is begun in the state on the left, then it evolves by the Schrödinger equation in some fixed finite time to the state on the right.

Then the Schrödinger equation (which is the principal law of QT) implies that measuring an electron initially prepared in a superposition yields:

$$\{|1>+|2>\}|r> \rightarrow |1>| \text{ reads '1'}>+|2>| \text{ reads '2'}>.$$

But the final state on the right is not an eigenstate of position for the pointer. So the orthodox interpretation of QT—more precisely the eigenvalue–eigenstate link—is committed to the pointer having no definite position.

There are clearly two main approaches for solving this problem. Either we somehow change the Schrödinger equation, so as to replace the above final state by an eigenstate of pointer-position. This approach is called 'collapsing the wave-packet'. I discuss it in section 5. Or we somehow supplement the eigenvalue–eigenstate link's meagre ascription of values: we postulate extra values. But in this chapter, I have no space to discuss this second approach. Suffice it to say that, like section 5's approach, it has versions that uphold physicalism and versions that violate it (see Butterfield 1995, pp. 145f; 1996).

4.3 Avoiding indefinite appearances

For our topic of QT and mind, it is also important to emphasize another contrast (briefly mentioned at the start of subsection 4.2). Namely, between:

- (DefMac): those strategies for solving the measurement problem that aim to secure a definite macrorealm, and so to explain why the macrorealm is as it appears to be; and
- (DefApp): those, perhaps more radical, strategies that allow an indefinite macrorealm and aim only to secure definite appearances (thus denying that it is as it appears to be).

We shall show in section 5 that this contrast cuts across the one at the end of section 4.2. That is, some versions of the 'collapsing-the-wave-packet' approach aim to secure a definite macrorealm whereas others aim only to secure definite appearances. (And, similarly, some versions of the 'extra values' approach aim for a definite macrorealm, such as the Bohm interpretation; whereas others aim only for definite appearances, such as the 'many minds' interpretation.)

To clarify the contrast between (DefMac) and (DefApp), it will be helpful to 'drive the measurement problem into the brain'. That is, it will be helpful to show how, according to QT's orthodox interpretation, a quantum-theoretic model of perception of the pointer will lead to an indefinite perception of pointer-position, in the case where the electron is initially prepared in a superposition. Since we never have such indefinite perceptions, orthodoxy will face a problem of 'indefinite appearances', just as much as one of 'indefinite macrorealm'. Showing this will occupy the rest of this subsection.

We can again consider a very simple model of perception, based on our toy-model of measuring an electron's momentum. Recall that we had:

$$\{|1> + |2>\}|r> \quad \rightarrow \quad |1>| \text{ reads '1'} > + |2>| \text{ reads '2'} >.$$

Now let us assume that the brain state corresponding to a person, Anna, observing the pointer's position, and believing it to be at '1 unit', is a quantum-state, call it | believes '1'>. And similarly for her brain state corresponding to observing and believing it to be '2 units': it is a quantum state, say I believes '2'>. In other words, let us assume a physical quantity for Anna's brain, with eigenvalues 1, 2, etc., whose eigenstates correspond to her beliefs in such a way that we can mnemonically call it 'belief-in-position'. (Despite the mnemonic, it is, of course, a normal physical quantity, perhaps some function of the positions, momenta, energies etc. of her brain's constituent particles.) And similarly for her brain state corresponding to her 'ready state', I alert> say.

These assumptions countenance using QT to describe the brain. That is a substantial contention, based of course on taking QT to be the fundamental theory of all (material, spatiotemporal) objects. (For further defence of these assumptions, see, for example, Butterfield 1995, pp. 146–7.) But given these assumptions, we can 'drive the measurement problem into the brain'. In terms of our toy-model, assuming that Anna is reliable on the eigenstates, in the sense

$$|1\rangle|r\rangle| \text{ alert}\rangle \ \rightarrow \ |1\rangle| \text{ reads '1'}\rangle| \text{ believes '1'}\rangle \quad \text{and}$$
$$|2\rangle|r\rangle| \text{ alert}\rangle \ \rightarrow \ |2\rangle| \text{ reads '2'}\rangle| \text{ believes '2'}\rangle,$$

the Schrödinger equation implies that measuring a superposition gives

$$\{|1\rangle+|2\rangle\}|r\rangle| \text{ alert}\rangle \ \rightarrow$$
$$\{|1\rangle| \text{ reads '1'}\rangle| \text{ believes '1'}\rangle+|2\rangle| \text{ reads '2'}\rangle| \text{ believes '2'}\rangle\},$$

which is not an eigenstate of pointer-position, nor of the quantity we called 'belief-in-position'. But Anna—we!—always have definite beliefs about pointer-positions. So (given the orthodox eigenvalue–eigenstate link), the fact that the final state is not an eigenstate of belief-in-position seems manifestly wrong.

To sum up: we have seen that QT faces a problem of 'indefinite appearances', just as much as one of 'indefinite macrorealm'. And there are two broad strategies for responding to these problems, namely (DefMac) and (DefApp) above. That is, we can either:

- (DefMac): somehow secure a definite macrorealm, and, more specifically, secure the successful predictions of classical physics, so that we can rely on a classical psychophysics for understanding the definiteness of appearances; or we can

- (DefApp): allow an indefinite macrorealm, and somehow secure only that appearances are definite.

In view of the discussion above, we expect that (DefApp) will involve some 'funny business' at the interface of brain and mind. And indeed, it is exactly here, under strategy (DefApp), that some proposals violate physicalism.

5   Collapsing the wave packet

In this section I discuss one approach to the measurement problem: the approach that postulates new physical laws to replace the Schrödinger equation. Such a change of the state is called 'the collapse of the wave-packet' (or 'state reduction'). More specifically, I take three versions of this approach. In subsection 5.1, I take that of Ghirardi, Rimini, Weber, Pearle *et al.*; and that of Penrose. In subsection 5.2 I discuss that of Wigner and Stapp. For all versions, I emphasize their consequences for brain or mind, and so for physicalism (I take them in increasing order of radicalism).

The first two versions (5.1) take the collapse of the wave-packet to be a purely physical process; and so aim to secure a definite macrorealm (strategy (DefMac) of subsection 4.3). Perhaps unsurprisingly, the new laws describing the collapse of the wave-packet seem entirely compatible with physicalism as defined in section 3. But these versions substantiate my first conclusion, announced in section 1: on these proposals, the way in which the mental supervenes on the physical would be very different from what most people, unversed in the controversies about QT, would expect. (The difference is more radical, for Penrose's version.)

On the other hand, Wigner and Stapp's version invokes mind to trigger the collapse of the wave-packet. This will yield my second conclusion: namely, this version gives a real-life example of the abstract problem confronting the definition of physicalism at the end of section 3. So here is an interpretation of QT that violates the spirit of physicalism, if not its letter.

5.1   Physical collapse

I will first discuss the proposals of such authors as Ghirardi, Rimini, Weber, Pearle and Percival (giving me the acronym GRWP), who propose precise equations for the collapse of the wave-packet (to one out of various alternative eigenstates). They do not aim to discuss the brain. But it turns out, surprisingly, that according to their proposals, the collapse sometimes happens, not in the external world but in the observer's nerves, such as in the retina. This is no threat to physicalism, but it *is* surprising.

I shall concentrate on a well-known precise proposal, called 'continuous spontaneous localization' ('CSL': Ghirardi *et al.* 1986; Pearle 1989). This postulates a jitter that continually gives 'little hits' to the quantum state, in addition to its usual evolution according to the Schrödinger equation. (These hits increase the system's energy: as mentioned at the end of subsection 4.1.) The jitter is a stochastic process, which in any individual case has a realization, with a prescribed probability. (Analogy: The jitter is like the probability space for 10 tosses of a coin; the realization is a specific sequence of 10 results, heads or tails.) Which realization happens in an individual case determines what happens to the system—what the final eigenstate is.

Any such proposed dynamics must somehow make the collapse mechanism ineffective in the microrealm (so as to recover the empirical success of QT's Schrödinger equation), but effective in the macrorealm. To do this, CSL chooses a rate of hitting and a size of hits so that the effect is almost always utterly minuscule for a microsystem. But on the other hand, each individual component of a composite system is subject to these rare and weak hits; and, when hit, it drags the other components with it. The result is that for a macrosystem with maybe 10^{23} components, the collapse is very fast (10^{-9} seconds). To take the time-honoured example: Schrödinger's cat is only superposed between life and death for a split second. This is surely acceptable, even to an advocate of the strategy (DefMac): for though it only gets the macrorealm to be definite at almost all times, nobody can be so certain that it is definite *always*.

The further details of CSL need not concern us. The mere fact that it links collapse to having a large number of components is enough to yield interesting consequences for physicalism. The point is best made by presenting an objection to CSL, due to Albert and Vaidman (see Albert 1992, pp. 100–11). The objection is wrong, but fruitful: for the reply brings surprises about the way in which the mental supervenes on the physical.

The idea of the objection is that some measurement results involve only a small number of microsystems. For example, a result may be registered by the arrival of a microsystem in one position rather than another on a fluorescent screen (like a TV screen). But its arrival excites only about 10 atoms, which then de-excite, each emitting a photon. Since retinal cells are so sensitive as to fire in response to a few photons, this is enough for a human to detect one result rather than another. So the objection is that since, at any stage in this process, only a few microsystems are involved, there will only very rarely be a GRWP collapse—contradicting the fact that our perceptions of the spots on the screen are always definite.

GRWP reply by analysing the physical process of a nerve cell firing. They show that the transport of ions involved in the firing requires sufficiently many particles being sufficiently well separated in space for a collapse to occur with overwhelming probability in, say, a hundredth of a second (see, for example, Ghirardi et al. 1995; subsection 5.2).

I should make two points about this reply. First, it involves no appeal to mind or consciousness: and so it poses no threat to physicalism. For it rests on the purely physical and contingent fact that, in the example, the nervous system is the first place where enough particles are involved for the theory to predict collapse. If our retinas responded only to millions of photons, or if we perceived a result only by reading words or by hearing (which both involve displacements of millions of atoms), all collapse relevant to our perception of a definite result would indeed occur outside the head.

This leads to my second point. Namely, that such examples of collapse occurring only in the head, so late in the causal chain of perception—although in no way threatening physicalism—are indeed surprising. For GRWP's proposals are examples of strategy (DefMac) of subsection 4.3: that is, of aiming to secure a definite macrorealm, which obeys classical physics (to a very good approximation). And on this strategy, one naturally expects psychophysics (which is to then secure the definiteness of appearances) not to involve any peculiarities of quantum theory. After all, consider the success of neuroscience in understanding perception, while using a wholly classical chemistry—a chemistry that models a molecule as like a group of billiard balls linked by rods, not subject to such quantum peculiarities as being in a superposition of two positions. (Nor a superposition of other quantities, including ones especially relevant to chemistry, such as handedness: classical chemistry does to countenance superpositions of left-handed and right-handed orientations of a molecule.) This example, and others like it, lead one to expect all such quantum peculiarities to 'die out' at the level of neuro-physiology, and so also at the level of psychophysics. (For a general discussion, bearing on reduction and cumulativism in science, see Rohrlich and Hardin 1983.) So it is surprising that GRWP's proposals entail that psychophysics involves such peculiarities.

To sum up: GRWP's proposals are compatible with physicalism, but hold some surprises about exactly how the mental supervenes on the physical.

I turn now to Penrose. I will argue that the same overall conclusion applies to him, though, in his case, the surprises for psychophysics are greater. Unlike GRWP, he is uncommitted about precise equations; but he thinks gravity is responsible for the collapse. He also deliberately aims to discuss the brain, indeed the mind. For he believes that although the collapse is a purely physical process, it involves non-computational physics; and that this physics will be relevant to brain action, because it will help explain consciousness, which Penrose believes to be non-algorithmic (non-comput-able). He even has proposals about how this physics operates in the microstructure of cells (1989, especially pp. 367–71; 1994, especially chapter 6, section 8 et seq., and chapter 7).

Unlike GRWP, Penrose does not yet have detailed models of how gravity induces collapse. But his idea is that the gravitational self-energy of the difference between two mass distributions considered to be in quantum superposition determines a rate at which collapse takes place to one of the two distributions (1994, sections 6.10–6.12, especially 6.12). This idea is akin, as

Penrose of course acknowledges, to GRWP's proposals; and this idea, suitably developed, seems quite as likely to solve the measurement problem satisfactorily, as are the proposals of GRWP.

I turn to Penrose's claim that consciousness is non-algorithmic. This claim is very controversial: it is largely based on an analysis of Gödel's monumental 1931 theorem about the incompleteness of arithmetic. As Penrose discusses (1994, part I), it is uncontroversial that Gödel's theorem establishes that human mathematicians are not using an algorithm that is both sound and knowably so. He goes on to argue that they are not using an 'unconscious algorithm'—that is, one that is sound but not knowably so. If that is right, then at least one aspect of human consciousness—namely mathematical understanding—would be non-algorithmic. Penrose goes on to suggest that other aspects of consciousness (with some other organisms included) are also non-algorithmic.

Penrose connects the idea that consciousness is non-algorithmic with collapse, by arguing that both classical physics and orthodox QT are, in the relevant senses, algorithmic. So he maintains that although collapse is a physical process happening all the time, outside the body, due to gravity (thus securing a definite macrorealm), it has a scientifically important role inside the brain: to supply the non-algorithmic physics that underlies consciousness—or at least mathematical understanding.

Penrose also proposes a biological locus for this non-algorithmic physics. After first conceding that neurones are 'too large' and 'classical' to be involved in this new physics (1994, sections 7.1–7.2), he proposes the microtubules that occur in cells. These tiny filaments have some promising features (1994, sections 7.3–7.7). In particular, they might allow an internal quantum state: (1) to be coherent, relatively unperturbed by the environment; and (2) to interact with a classical computation performed along the surface of the microtubule, by the varying configurations of the tubulin protein molecules making up that surface (rather in the manner of cellular automata).

Turning to assessment, I obviously cannot do justice to this slate of imaginative, and mutually connected, proposals. It must suffice to make two comments. First, my own reactions, for what they are worth. I am happy to allow that gravity is crucial to collapse of the wave-packet, if such there be; and also that consciousness, or at least mathematical understanding, is non-algorithmic. But I am not convinced that both classical physics and orthodox QT are, in the relevant sense, algorithmic. So collapse may not be the only mechanism that could provide the non-algorithmic basis of mathematical understanding. And so microtubules may not be the only place to look for such a basis.[9]

Second, the overall conclusion for us is as it was in the case of GRWP. If Penrose's proposals are true, then physicalism is intact, but psychophysics is full of surprises. As Penrose says: 'it is only the arrogance of our present age that leads so many to believe that we now know all the basic principles that can underlie all the subtleties of biological action' (1994, p. 373).

5.2 Mental collapse

I turn to the proposal of Wigner and, more recently, Stapp that mind (or consciousness) itself produces the collapse of the wave-packet; which yields my second conclusion, as announced in section 1.

The idea is that the usual Schrödinger evolution holds throughout the physical realm, and is broken only at the interface of brain and mind. Once the mind sees one result (in my example, once Anna sees one pointer-position), the superposition is replaced by an eigenstate—namely the one corresponding to the result seen.

The first point to notice about this proposal is that it is a hybrid of the broad strategies (DefMac) and (DefApp) in subsection 4.4. For once consciousness has 'done its stuff' and collapsed the wave-packet, the macrorealm really is definite (in the quantities of which the collapsed state is an eigenstate). But then the usual Schrödinger evolution takes over again. So in general, there is no guarantee that the macrorealm stays definite (in those favoured quantities) as time goes on; in particular, when consciousness stops looking—'when there's no one about in the Quad'.

Among the founding fathers of QT, Wigner (1962) expresses this proposal most clearly (although he later changed his mind, and von Neumann and Heisenberg expressed similar views). Stapp (1993) revives the proposal; of course acknowledging these precursors. Unsurprisingly, the authors differ about what exactly it takes to reduce the state. Thus Heisenberg allows collapses to occur not only in humans, but also in cats (and even in inanimate macroscopic apparatuses). Stapp (1993) joins him in this, but has now withdrawn this allowance, on grounds of parsimony (1995).

There are two obvious questions confronting this proposal, as so far stated. I need to pose them, but not to answer them. For posing them will be enough to yield my conclusion: that this proposal violates the spirit, if not the letter, of physicalism as I formulated it in section 3.

The first question is: with which quantity's eigenstates does consciousness replace the initial superposition? No doubt, the basic idea of the answer must be 'the quantity that seems to have definite values'. But then the question is: what general laws, if any, constrain which quantity that is? If there are such laws, can they be expressed in wholly physical terms; or is there some irreducible invocation of mind? At first sight, it seems that (1) if the laws are physical, then physicalism might yet be upheld on the Wigner–Stapp proposal; whereas (2) if they must invoke mind, then it cannot be upheld.

The second claim (2) is straightforward: no doubt, the existence of irreducibly mental laws implies that physicalism is false. But about the first claim, (1), matters are not so clear. For we are back at the problem that confronted section 3.4's formulation of physicalism (i.e. using choice 2: supervenience across the set of nomically possible worlds). That problem was: if mental properties are correlated with physical properties according to strict laws, then this formulation can be true, even though, intuitively, physicalism is false. (In other words, choice 2 seems too weak, and the range of worlds across which physicalism claims supervenience needs adjusting.)

This problem applies here too. Even if the laws (governing which quantity's eigenstates are collapsed onto) are expressed in wholly physical terms, the fact that only with consciousness is there any collapse means that, intuitively, physicalism is false. Similarly, if there are no laws at all about which eigenstates the collapse is onto: intuitively, physicalism is false.

A similar point applies about the second obvious question confronting this proposal. This question concerns the time of collapse. Thus, we can ask: what if anything constrains or determines when the collapse occurs? Suppose that laws do so, and they can be expressed in wholly physical terms. Then nevertheless, just as before, intuitively physicalism is false—because only with consciousness is there any collapse. Similarly, if there are no laws at all about when collapse occurs: intuitively, physicalism is false.

I cannot here pursue how one might answer these two questions, nor how to improve the formulations of physicalism. Suffice it to say, by way of summary: here is a real-life, albeit rather undeveloped, proposal, which violates the intuitive idea of physicalism—and provides an example of a problem that is discussed only in the abstract in philosophy. One could therefore try to use the proposal as a test-case for conjectured precise formulations of physicalism.

Acknowledgements

I thank Peter Lipton for comments on a draft of this paper; participants at the conference for discussions; Chris Daly, Sue James, and Tim Crane for advice on references; Bay Whitaker for making a transcript; John Cornwell, Rebecca Beasley and the Science and Human Dimension Project at Jesus College, Cambridge for the invitation—and for patience; and David Lewis for the inspiration of his metaphysical system, in writing sections 2 and 3.

Notes

1. In any epoch since the time of Galileo, there have, of course, been processes or phenomena that seemed not to admit mechanical explanations, and for which some scientists accordingly hypothesized a non-mechanical explanation. But the definitive demise of mechanism within physics came as a result of the success of clearly non-mechanical theories; namely, theories of electromagnetism, around 1900.
2. Furthermore, in section 3 we will see that my formulation of physicalism dovetails neatly with this strategy of 'looking widely'.
3. Though these states are hard to describe, for the reasons given, they can to some extent be conveyed to others—witness the works of the impressionists, and writings about the stream of consciousness, by authors such as Proust, Joyce, and Woolf.
4. For the prevalence, and legitimacy, of philosophy adding detail and precision to claims that are widespread in the contemporary culture, see Craig (1987). He gives telling case studies, drawn from throughout the last 400 years.
5. Teller (1984) says so explicitly; others are less explicit—for example, McGinn (1991). But McGinn has recently (1996) been explicit in his criticism of Chalmers's (1995) claim that 'zombies'—physical duplicates of sentient beings like you and me, but lacking sentience—are logically possible.
6. But I agree with Lipton that Searle goes wrong. Searle's analogy for making such a coextension credible—namely, the relation of solidity to lattice-vibrations (1992, pp. 112–26; Chapter 2 this volume, replies to theses 7 and 8) has no relevant difference from examples, such as heat and molecular motion, which Searle claims to be false analogies.
7. Indeed, I believe that a logically stronger doctrine about the mind–matter relation—namely, in the jargon, analytical functionalism combined with contingent type–type mind–brain identity theory—is plausible; and can accommodate these features of consciousness.
8. Agreed, to a philosopher of probability, this identification of a fact (having a value) with its holding with probability 1, will seem a howler. And as we shall see, it may well be wrong.
9. For a recent exchange emphasizing biological details, see Grush and Churchland (1995) and Penrose and Hameroff (1995).

References

Albert, D. (1992) *Quantum mechanics and experience.* Harvard University Press, Cambridge, Mass.

Baker, G. and Morris, K. (1993). Descartes unlocked. *British Journal for the History of Philosophy,* **1**, 5–27.

Burge, T. (1979). Individualism and the mental. In *Midwest Studies in Philosophy,* Vol. IV (ed. P. A. French *et al.*). University of Minnesota Press, Minneapolis.

Butterfield, J. (1995). Worlds, minds and quanta. *Aristotelian Society Supplementary Volume,* **69**, 113–58.

Butterfield, J. (1996) Whither the minds? *British Journal for the Philosophy of Science,* **47**, 200–21.

Chalmers, D. (1995). *The conscious mind.* Oxford University Press, Oxford.

Craig, E. (1987). *The mind of God and the works of man.* Oxford University Press, Oxford.

Crane, T. (1995). The mental causation debate. *Aristotelian Society Supplementary Volume,* **69**, 211–36.

Crane, T. and Mellor, D. (1990). There is no question of physicalism. *Mind,* **99**, 185–206; reprinted in Mellor's *Matters of metaphysics* (1991), Cambridge University Press, Cambridge.

Dennett, D. (1991). *Consciousness explained.* Penguin, London.

Dupré, J. (1993). *The disorder of things.* Harvard University Press, Cambridge, MA.

Ghirardi, G., Rimini, A., and Weber, T. (1986) Unified dynamics for microscopic and macroscopic systems. *Physical Review,* **D34**, 470–91.

Ghirardi, G., Grassi, R., and Benatti, F. (1995). Describing the macroscopic world: closing the circle in the dynamical reduction program. *Foundations of Physics,* **25**, 5–40.

Grush, R. and Churchland, P. (1995). Gaps in Penrose's toilings. *Journal of Consciousness Studies,* **2**, 10–29.

Healey, R. (1978). Physicalist imperialism. *Proceedings of the Aristotelian Society,* **74**, 191–211.

Horgan, T. (1982). Supervenience and microphysics. *Pacific Philosophical Quarterly,* **63**, 29–43.

Kim, J. (1984). Concepts of supervenience. *Philosophy and Phenomenological Research,* **45**, 153–76; reprinted in Kim's *Supervenience and mind* (1993), Cambridge University Press, Cambridge.

Kripke, S. (1979). A puzzle about belief. In *Meaning and use* (ed. A. Margalit), pp. 230–75. Reidel, Dordrecht.

Kripke, S. (1980). *Naming and necessity.* Blackwell, Oxford.

Leggett, A. (1987). *The problems of physics.* Oxford University Press, Oxford.

Lewis, D. (1983). New work for a theory of universals. *Australasian Journal of Philosophy,* **61**, 343–77.

Lewis, D. (1986a). *Philosophical Papers,* Vol. 2. Oxford University Press, Oxford.

Lewis, D. (1986b). *On the plurality of worlds.* Blackwell, Oxford.

Lewis, D. (1990). What experience teaches. In *Mind and cognition* (ed. W. Lycan), pp. 499–519. Blackwell, Oxford.

Locke, J. (1972). *Essay concerning human understanding.* Dent Everyman, London. (Originally published 1690.)

McGinn, C. (1991). *The problem of consciousness.* Blackwell, Oxford.

McGinn, C. (1996). Review of *The conscious mind* by D. Chalmers. *Times Higher Education Supplement,* 5 April, vii–ix.

Monod, J. (1972). *Chance and necessity.* Collins, London.

Oppenheim, P. and Putnam, H. (1958). The unity of science as a working hypothesis. In *Minnesota studies in philosophy of science,* Vol. 2 (ed. H. Feigl, M. Scriven, and G. Maxwell), pp. 3–36. University of Minnesota Press, Minneapolis.

Pearle, P. (1989). Combining stochastic dynamical state-vector reduction with spontaneous localization. *Physical Review,* **A39**, 2277–92.

Penrose, R. (1989). *The emperor's new mind.* Oxford University Press, Oxford.

Penrose, R. (1994). *Shadows of the mind.* Oxford University Press, Oxford.

Penrose, R. and Hameroff, S. (1995). What 'gaps'? *Journal of Consciousness Studies,* **2**, 99–112.

Putnam, H. (1975). The meaning of 'meaning'. In *Mind, language and reality,* pp. 215–71. Cambridge University Press, Cambridge.

Rohrlich, F. and Hardin, C. (1983). Established theories. *Philosophy of Science*, **50**, 603–17.

Searle, J. (1992). *The rediscovery of mind*. MIT Press, Cambridge, Mass.

Stapp, H. (1993). *Mind, matter and quantum mechanics*. Springer-Verlag, New York.

Stapp, H. (1995). The integration of mind into physics. In *Fundamental problems in quantum theory* (ed. D. Greenberger and A. Zeilinger), pp. 822–33. Annals of New York Academy of Sciences, New York.

Teller, P. (1984). A poor man's guide to supervenience and determination. *Southern Journal of Philosophy*, Suppl. **22**, 137–62.

Wigner, E. (1962). Remarks on the mind–body problem. In *The scientist speculates* (ed. I. J. Good), pp. 284–301. Heinemann, London.

CHAPTER 8

Putting ourselves together again

MARY MIDGLEY

Something called the 'problem of consciousness' is now beginning to worry scholars in several disciplines, including some in which, until lately, such words were not supposed to be heard at all. Territorial disputes are even breaking out over whether this new problem is the property of scientists or philosophers.[1] If we mind about this debate, we can probably divide the spoils to some extent, because consciousness actually raises many problems, not just one. But the most interesting puzzles are more or less bound to have both scientific and philosophical aspects. So perhaps we shall have to try and co-operate, hard though we may find it. And, in what looks like the central and most difficult puzzle, both these aspects are surely present.

That central puzzle is not, I think, 'how consciousness evolved', nor is it 'how we would know it was there if we didn't happen to be aware of it already'. These questions are both secondary to what most worries people. That central worry is: how can we rationally speak of our inner experience at all? How can we regard our inner world—the world of our everyday experience—as somehow forming part of the larger, public world which has been described in terms that seem to leave no room for it? On what map can both these areas be shown and intelligibly related? This is a genuinely difficult issue, not a false alarm. It will not yield simply to familiar methods and a good injection of research money. But it is not desperate either. We can do something about it if we are willing to stand back, to look at things from farther off and to admit the size of the question. Current panic about consciousness arises, I believe, from trying to treat it as if it were a much smaller issue than it is.

The size of the problem

The analogy that I would like to suggest here comes from geography. We are not looking for the relation between two places on the same map. We are trying to understand the relation between two maps of different kinds. And that is a different sort of enterprise. At the beginning of an atlas, we usually find a number of maps of the world. Mine gives, for instance: World Physiography (Structure and Seismology); World Climatology (Mean Annual Precipitation; Climatic Fronts and Atmospheric Pressure); World Vegetation; World Political; World Energy; World Food; World Air Routes, and a good many more. If we want to understand how this bewildering range of maps works, we do not need to pick on one of them as 'fundamental'. We do not need to find a single atomic structure belonging to that one map and reduce all the other patterns to it. We do not, in fact, have to do once more the atomizing work that has already been done by physics. Nor will that work help us here.

What we do need is something different. We have to relate all these patterns in a way that shows why all these various maps are needed, why they are not just contradicting one another, why they do not just represent different alternative worlds. To grasp this, we always draw back to consider a wider whole. We look at the general context of thought within which the different pictures arise. We have to see the different maps as answering different kinds of question—questions that arise from different angles in different contexts. But all these questions are still about a single world, a large world that can be rightly described in all these different ways and many more. That background—and not a common atomic structure—is what makes it possible to hold all the maps together. The plurality that results is still perfectly rational: it does not drop us into anarchy or chaos.

When we are using an atlas, we can do this quite easily so long as we use each of the maps as a whole, because the same coastlines appear on all of them. This helps us to relate the various pictures to the world they represent. If, however, we forgot that wider context and tried to examine a smaller area on its own, we would be in trouble. For instance, if someone decided to investigate a particular square of Central Africa or Australia by cutting out and magnifying the parts of all these maps that showed that area, the lines on many of these squares would not seem to bear any meaningful relation to one another at all. The Political map, especially, might just show a single straight line running right across on the whole area—perhaps pink on one side, blue on the other—something quite extraordinary that no other map ever shows. (There are no straight lines in nature.) And the Airline map might be rather similar, but showing a different line.

At this point, anyone who had been using this method might give up. They might say that their investigation had shown that the maps disagreed so badly that there could only be one of them that was correct and fundamental—only one map really showing the world at all; namely, the one that they had backed in the first place. And this, it seems to me, is very much the way in which many people are now trying to investigate the problem of consciousness.

When we try to consider that problem our project (as I have suggested) is not to relate two things that already appear on the same conceptual map. It is about how to relate two maps that answer questions arising from different angles. Consciousness is not just one object, nor one state or function of objects, among others in the world. It is not (as people often suggest) something roughly parallel to digestion or perspiration. It is the condition of a *subject*, someone for whom all those objects are objects. The questions it raises are therefore primarily about the nature of a person as a whole, a person who is both subject and object.

When we ask how consciousness can be a feature of the world, we are asking how we ourselves—as subjects—can be both items in the world and aware of it as a whole? This is not a factual question. It is a question about how we can best think about an item that has this double position. How does that awareness fit, conceptually, both into the world and into the rest of our complex nature? Science itself has so far depended on and nurtured realism about the world. The problem now is, how to be realist about subjects, which are, it should be stressed, a natural phenomenon, not some kind of invented spooks.

Consciousness, then, is not just one more phenomenon. It is the scene of all phenomena. It is the place where appearances appear. It is the viewpoint from which all objects are seen as objects. The first questions that arise about it are questions about ourselves. These questions have to come before more strictly scientific questions about the place of consciousness in the outer world, such as how it evolved. And it is important that this 'consciousness' contains a whole mass of more dynamic items besides simple appearances. It contains things such as emotions, efforts, conflicts, desires—aspects of our active participation in what goes on around us. If we once start to sort out these things, we shall probably have to think about agency and free-will as well as perception (and—if one may mention it—it is time we did so).

Prior taboos and their decay

All this is very hard to deal with today. Current scientific concepts are not adapted to focusing on the topic. Indeed, many of them have been carefully adapted to exclude it, much like cameras with a colour filter. People have not, indeed, usually believed that they were unconscious. The fathers of modern science took their consciousness for granted, since they were using it to practise science. Even the Behaviourists, who did try to deny this awkward fact, still tacitly presupposed it. They assumed (for instance) that it was still worth their while to talk, write, and read about scientific subjects. Although they were officially epiphenomenalists, they took it for granted that their thoughts would actually affect their writings and thereby have an influence on the world. They assumed that conscious beings were there to receive and understand their words. They assumed, too, that they had useful colleagues —that testimony from other conscious beings was a sound source of scientific knowledge. And so forth. This background community of subjects has to be presupposed if any form of connected thought is to be possible at all.

Until very lately, however, these conscious subjects were not seen as something that science itself could study. They had been shut out of its domain, with good reason, during the Renaissance. Galileo and Descartes saw how badly the study of objects had been distorted by people who treated these objects as subjects, people who credited things like stones with human purpose and striving. So they ruled that physical science must be objective. And this quickly came to mean, not just that scientists must be fair, but that they should treat everything they studied only as a surd, passive, insentient object.

We know that this abstraction made possible three centuries of tremendous scientific advance about objects. Today, however, this advance has itself led to a point where consciousness has again to be considered. Enquiries are running against the limits of this narrowed focus. In many areas, the advantages of ignoring ourselves have run out. This has happened most notoriously in quantum mechanics, where physicists have begun to use the idea of an observer quite freely as a causal factor in the events they study. Whether or not this is the best way to interpret quantum phenomena, that development is bound to make people ask what sort of an entity an observer is, since Occam's Razor has so far failed to get rid of it. This disturbance, however, is only one symptom of a growing pressure on the supposedly subject-proof barrier, a pressure that is due to real growth in all the studies that lie close to it.

The pressure is naturally strongest in the social sciences, especially in psychology. As a direct consequence of the success of physical science, the social sciences were initially designed to imitate its methods as far as possible. Sometimes this has worked well, but often it has not. Social investigators who have tried to confine themselves to the methods used by their physical colleagues have repeatedly run into trouble at certain points, where they found that they simply couldn't make progress without considering their subjects as subjects—their people as people. At these points, not even the most objective observer could dismiss the subjects' own subjective point of view as irrelevant. It had to be acknowledged that people were in some ways different from stones.

Early in the twentieth century, however, behaviourist psychologists built their defensive rampart against this demand. They ruled that consciousness either did not exist at all or, if it did exist, didn't matter. It need not be studied because it was inert; it had no consequences. Outward behaviour was self-contained and could be studied on its own.

This highly ascetic programme was never very successful. It turned out not to be possible to describe people's behaviour without using terms that referred to their subjective points of view. The whole vocabulary of useful words was slanted to take in that viewpoint. Yet for a long time it seemed that the behaviourist project was the only possible way in which psychology could make good its status as a science. Behaviourist ideology was thus so fervently launched and so fiercely policed that its principles prevailed for the best part of a century, long outlasting the detailed work that was supposed to support them. During that time, it was as much as a social scientist's career was worth to be caught taking what B. F. Skinner so oddly called 'an anthropomorphic view of man'.[2] The methodological argument that underlay this view ran much like this:

> Only what science studies is real,
> Science cannot study consciousness,
> So: Consciousness is not real.

Although the conclusion has officially been abandoned, both these premises are still widely accepted. Hence much of our present difficulty.

As I just mentioned, the behaviourist taboo on ordinary ways of thinking has lately been dying out because of its own failures. The church of academic orthodoxy now once more lets its members see these problems. But if we want to see them clearly, we are going to need a more suitable set of concepts. The terms that served seventeenth-century thinkers for dismissing conscious subjects from scientific attention are not likely to be the best ones for bringing them back into focus now.

The return of the first person

We need other ways of thinking. We have to stop thinking of consciousness as a peculiar, isolated feature of certain objects—as just one particular state or function of certain organisms—and start to think of it rather as a whole point of view, equal in size and importance to the objective point of view as a whole. And we shall not get far with this if we start our investigation by worrying about whether we can recognize other people's consciousness—about the so-called 'problem of other minds'. To suppose that we have a problem about the existence of other minds is to be in trouble already. It is to have started in the wrong place—Descartes's wrong place. If we once sit down in that place, we shall never get rid of the problem. (Bertrand Russell, who was wedded to this starting-point, never did get rid of it.) This approach conceives of minds, or consciousnesses, unrealistically as self-contained, isolated both from each other and from the world around them. It is terminally solipsistic.

To avoid this unreal isolation, we had better attend in the first place to the examples that are most familiar to us—namely, our own experience and how it interacts with that of others whom we already know well. Consciousness is not something rare and exotic found only in experimental subjects or in scientific observers. Nor does it only show us certain odd phenomena such as colours and dreams and hallucinations. It is not primarily an observation-station. It is the crowded scene of our daily lives. And the main dramas going on in it do not concern just observation or perception but quite complex, dynamic currents of feeling and efforts to act. If we mean to do justice to this complexity, we shall have to take seriously the rich, well-organized language that we use about it every day. That language does not just express an amateur 'folk psychology'. It is the indispensable working skeleton of all our thought, including, of course, our thought about science.

The starting-point; selves in solitary confinement

Perhaps this will become clearer if we look back again at the unreality of the traditional scheme that is now making this problem so hard. That scheme was, of course, Descartes's sharp division of mind from body. To make the natural world safe for physics, Descartes pushed consciousness right out of it into a separate spiritual world, treating each soul or mind as a spiritual substance, made of a stuff alien to other earthly items.[3] Today, we cannot possibly put this extra entity, this disembodied mind back into the natural world, though Descartes's dualistic followers still try. (The suspicion that we may have to restore it may account for some of the alarm that now surrounds this topic.) The trouble is not just that physics leaves no room for this kind of entity, but that thinking creatures could not possibly be isolated entities of this kind. Thought involves communication. Cartesian beings, isolated in their separate shells of alien matter, could never even have discovered each other's existence. What thinks has to be the whole person, living in a public world.

The 'problem of other minds' arises from positing this solipsistic self, which we might call Descartes's Diamond. It is a hard, impenetrable but very precious isolated sentient substance, which sits at its console in its windowless tower communicating with other, similarly secluded diamonds by signals run up between towers and relayed to each inhabitant by a perpetual miracle. Real people, by contrast, are embodied beings living in a public world.

Descartes's reason for introducing this awkward kind of soul was his need to compromise between three conflicting kinds of demand—the position of traditional Christianity, the need to segregate physics from other studies, and the demands of everyday language. The Diamond never fitted any of these systems at all well and as time has gone on the misfit has grown steadily worse. It was always an unstable notion. When it was taken seriously, it has usually tended to expand into absolute idealism—the idea that spiritual diamond-stuff was actually the stuff of the whole universe, a stuff that underlay physical matter as well as souls. Hence the mentalist tradition that runs through Leibniz, Berkeley, Hume, and Hegel to modern phenomenalism.

Today, by contrast, many educated people, and in particular many scientists, assume that the opposite metaphysic—materialism—has simply conquered this whole trend and now reigns unchallenged. But it seems increasingly clear that one extreme view is no more workable than the other. That is why the unstable notion of an isolated self has not gone away. The ghost still haunts the machine because the machine has not changed thoroughly enough to do without it.

Solitary pseudo-souls have been a surprisingly persistent image, often welcome to the individualism of our age. These chips off Descartes's Diamond can be found sticking in the works of many supposedly secular conceptual schemes. For instance, there is existentialist ethics, where the Will, which is solitary and independent of the entire natural constitution, is alone considered to be the authentic person.[4] Such doctrines disconnect the conscious self entirely from the world around it. There are also others, popular today, which do allow for a social context but deny a bodily one. In sociology for instance, there is still quite a widespread belief that human behaviour can only have social causes, not biological ones, so that the constitution of people's bodies can have no effect on their personality. And, again, there is post-structuralist literary criticism, in which the arbitrary relation between Sign and Signified is supposed to cut literature off from the everyday world that it might seem to be about, leaving it in some kind of spiritual mid-air as a purely cerebral network of texts. Most remarkable of all, however, there are the electronic dreams of downloading human conscious-ness onto computers, dreams which take it for granted that the personality is a kind of software that does not need the body because it can be run with equal ease on any kind of hardware. It may, then, even be sent ambitiously off to outer space, there to achieve eternal enlightenment, freed from the limitations of the earth.[5] But why, we might ask, should earthly creatures go so far in order to do that?

Does the fish soar to find the ocean?
The eagle plunge to find the air?
That we ask of the stars in motion
If they have rumour of thee there?

Not where the wheeling systems darken
And our benumbed conceiving soars!—
The drift of pinions, would we hearken
Beats at our own clay-shuttered doors.

The angels keep their ancient places;—
Turn but a stone and start a wing!
'Tis ye, 'tis your estranged faces,
That miss the many splendoured thing.[6]

From the materialistic side, too, there is the medical model that represents patients simply as bodies sent to be chemically and physically repaired, on the assumption that their conscious life is a separate matter irrelevant to that process.

Living in the world

I cannot spend more time here on these doctrines, some of which I have discussed elsewhere.[7] I think it is better now to do the more positive thing, which should always be done when one attacks a dominant image, and to try, however crudely, to suggest how we might think instead. What kind of image might we use, then, as a corrective for this strange isolationism? Well, we could do worse than listen for a start to A. E. Housman's view of the matter:

From far, from eve and morning
  And yon twelve-winded sky,
The stuff of life to knit me
  Blew hither; here am I.

Now—for a while I tarry
  Nor yet disperse apart—
Take my hand, quick, and tell me
  What have you in your heart?

Speak now, and I will answer
  How can I help you, say;
Ere to the wind's twelve quarters
  I take my endless way.[8]

Now I am by no means insisting on Housman's whole worldview. There may well be parts of that that we do not want at all. But this poem surely has two striking features that are right:

1. This 'I', this subject of Housman's, is as far as possible from being a pure, isolated diamond. It is a composite being. It has been formed by and out of the natural world. It is still a part of that world and wholly dependent on it.

2. It is continuous with that outside world in being essentially social—not solitary. This self never started life alone. It does not have any problem of other minds. It asks someone flatly (as any of us might) 'what's on your mind?, what have you in your heart?'. It speaks directly to another, with whom it shares a language and for whom it is deeply concerned. It is, in fact, a genuine member of our species.

The importance of babies

At this point, we might do well to remember that this is a species whose members, as babies, communicate with other people long before they try to handle inanimate objects. They also learn other people's names before they learn their own names. And they learn to talk about other people's mental states long before they become introspective enough to discuss their own. From birth, they are equipped—just as other young social animals are—with the right capacities for expressive behaviour, and with the power to interpret that behaviour in others. These capacities are not imitated from others later. Babies born blind smile, and non-blind babies can at once interpret smiles. Deaf babies cry. And so forth.[9] This innate repertoire makes it possible for human babies, just like other primate babies, to start communicating as soon as they begin to be aware of the world at all and long before they take any other sort of action.

That is how human infants manage to take in, quite directly, the mass of facts about other people's attitudes that will be the foundation of all their social knowledge and that will—among other important things—also make it possible for them to learn to talk. They do not need inference to do this. In fact, in our species, any social inference there may be is primarily directed inwards, from our knowledge of others to ourselves, rather than outwards by analogy from ourselves to them. We gradually learn to apply to ourselves the words that we already use to describe other people's moods and characters.

We can indeed, then, have a problem of self-knowledge—a 'problem of one's own mind'. We are often amazingly ignorant about our own condition and our own motives. This ignorance can become important later in life and when we are more mature it is our business to deal with it. But there is no original problem of other people's minds. For us humans at least (whatever may happen elsewhere in the universe) the whole use of language depends on, and arises out of, the deep, innate, unshiftable sense that we are among others who experience the world in roughly the same sort of way that we do ourselves. This sense is constitutive of our thinking, not a dispensable part of its content. Autistic people, who apparently lack that sense, tend to have great difficulty in using language at all.

Do we need proof that we are not alone?

Do we need some sort of proof for very general assumptions such as the assumption that we live in a public world? If so, one might ask, for a start, whether it is this assumption that needs proof or its opposite? If someone decides to assume that he does indeed live alone and has invented the companions that he has so far believed to surround him, would that assumption be less in need of defence than assuming a shared world or more so? Would it somehow be more economical?

We certainly do not know enough about the initial conditions to say which of these situations would be abstractly the more probable in an imaginary world. But in considering the world that we do know we are better off. The way to test such assumptions is to ask which of them makes more sense in the context of that world. The only kind of proof that they can have consists in showing that they are necessary to make thought and language possible.

We can see this well over two familiar assumptions:

1. That nature is regular or lawful—that unobserved facts will go on being like the observed ones so that the future will be like the past.
2. That in general we can trust our faculties—that our senses, memory and reasoning powers are not misleading us radically all the time.

Of course sceptics are right in saying that propositions like these are too wide to be checked in experience. No check could conceivably be adequate. They sometimes conclude from this that our belief in these assumptions is irrational. Indeed, they sometimes hint that if we were less lazy we would not believe them. Thus Hume, rejecting the reality of causation, concluded that 'if we believe that fire warms, or water refreshes, it is only because it costs us too much pains to think otherwise'.[10]

But this sets a quite unreal standard of rationality. The objection to dropping these basic assumptions is not just that thought becomes hard without them but that it stops altogether. Hume makes it sound as if we could go on on this path if we tried, just as we could, with an effort, dismiss a particular belief that we found to be prejudiced and unfounded. That kind of limited dismissal, however, leaves the whole background of our ordinary beliefs still standing. It still leaves us a world. By contrast, the notion of dismissing that whole background—of losing the basic conditions that make any experience reliable at all—produces a total conceptual vacuum in which the dismissal itself would lose all meaning along with everything else. There is no conceivable point to which thought would then move. Attempts at disbelief of this kind would not just run into emotional difficulties due to laziness. They would hit a logical block, like attempts to square the circle.

This cannot be what rationality requires. To the contrary, rationality demands that we should accept the conditions which are evidently necessary for reasoning. It does not order us to reject them in a desperate attempt to get an irrelevant sort of proof. There is nothing fishy about this. The word *proof* itself simply means test, and different kinds of test are needed for propositions that do different kinds of work.

If this general point about what is rational and what is not—what needs empirical proof and what does not—is clear, then we can see how it bears on our present argument. The fact that each of us is not alone in the world, that we live among others sufficiently like us to communicate with us, is one more of these basic conditions needed to make human thought possible at all. It is unlucky that Descartes failed to see this when he shaped his systematic doubt. That is how he came to forge the unreal, solipsistic conception of the mind that we have been discussing.

Descartes's oversight is remarkable in view of the stress he laid on the fact that thought needs language. For language is clearly a corporate, social phenomenon. Anyone who is in a position to say, as Descartes suggested, 'I think, therefore I am' has to be heir to a rich and widely shared linguistic tradition and must, therefore, be a member of a widespread company of similar beings. More generally still, all intelligent animals are social animals. Their thought is always a set of tools forged by a whole community. But for people in particular, the elaboration of these tools—the richness of a language that requires much time and trouble to learn—makes the idea of a solipsistic life wholly inconceivable.

Dignity, agency, and fatalism

These two propositions—physical and social continuity with the surrounding world—are not really contentious. They are obvious and fundamental truths about human subjects. In practice we all accept them. But many theorists have resisted admitting them, because they felt that this biological and social dependence on the world weakened the subject's dignity intolerably. And it is quite true that our dependence on the world does mean that we cannot have the guaranteed metaphysical autonomy that Descartes's Diamond was supposed to enjoy.

In spite of that dependence, however, we see that Housman's 'I' also still manages to function vigorously as an active agent. Although this 'I' knows that it is composite and transient, a weak, shaky fallible thing at the mercy of the natural world, it has no hesitation in offering to act. And there is no reason why it should have any such hesitation. It asks what to do, and this is a real question. It means to try to do it.

This self is not held back from making that offer by any theoretical fatalism, any conviction that it is really just a helpless, anonymous cog in the cosmic machine. (Housman, like the Stoics, was fatalistic about outside conditions, but never about internal freedom.) And as we all know, there is absolutely no reason why it should be held back by that kind of fatalism. Someone who is confronted with a friend who needs help is not likely to reply that, since we live in a deterministic world, he can't actually do anything because he is unfortunately unable to act freely. And if anybody did make that reply, the right name for it would not be 'scientific correctness' but 'humbug'.

Two sides to our head?

This two-sidedness of our nature certainly is very surprising. We might celebrate it with some more poetry, this time from Kipling:

> Much I owe to the Lands that grew,
>   More to the Lives that fed,
> But most to Allah, Who gave me two
>   Separate sides to my head

> I would go without shirt and shoes
>   Friends, tobacco and bread
> Rather than for an instant lose
>   Either side of my head.[11]

Like all good poems, this one uses rich imagery that can seem to point towards many different kinds of two-sidedness. It certainly does suggest the familiar division between thought and feeling, which has a good dealt to do with our present topic. But the kind of two-sidedness that we're now discussing—the combination of continuity with the world and independent action—seems to me to be near the centre of these many dualisms. This paradoxical combination is surely as essential to our life as Kipling suggests. yet it has always proved very hard to do justice to it in theory.

How are we to remember that we are, on the one hand, earthly organisms, animals operating within a physical pattern, a pattern that we can treat as fixed and predestined when we are merely observers taking the outside viewpoint, but that we are also, on the other hand, agents, beings who not only can but must choose what to do? When we stand at our inner viewpoint we are constantly forced to take decisions. This is the freedom of practical thinking, a freedom that we need, not only to act but to think as well. These practical choices cannot possibly be reduced (as Skinner sometimes tried to reduce them) to simply predicting what we might be expected to think or to do.[12] That is something quite different.

It is no wonder that we have evolved two rather distinct ways of talking for these two different contexts. Much of the time we can use these two languages side by side without noticing any clash between them. As Kant said, we use the deterministic one most in speculative thinking, especially when we think we are doing something called 'science', and the active one most for practical life. Often we can split ourselves neatly in two so as to duck possible contradictions between them. But we cannot do this all the time. And if we are interested in the larger scene—if we want to put the whole jigsaw together —we cannot avoid the problem of how to relate these two modes. Above all, we have to relate them when we think about personal identity—about what is, and what is not, essential in our lives, about the kind of being that each of us is as a whole.

Having it both ways

In recent times, however, scholars have not tried very hard to take that comprehensive viewpoint because they have preferred to make war. This war has usually been seen as being about which stuff the world is made of, rather on the Pre-Socratic model. It was assumed to be made of one kind of stuff, either of mental diamond-stuff or of matter. Materialists and Idealists shot arguments at each other to decide which stuff should prevail and which language could be hailed as describing the real world. The other language could then be sidelined as just something provisional, something handy for ignorant people but mildly misleading—in fact, a folk language. This inferior jargon might then still be used in everyday life until a better code was invented. But we were urged to remember not to take its concepts too seriously. As Bishop Berkeley kindly put it, we could 'think with the learned and speak with the vulgar'.[13] And at present 'thinking with the learned' is expected to mean thinking as a devout materialist, exclusively in terms of insentient objects.

The reason why this way of splitting our lives cannot work is that the vulgar, the folk, are ourselves, not poor relations whom our learned selves can disown. They constitute the core of our being. The learned ways of thinking and talking that we develop for certain limited purposes are secondary extras. These techniques cannot displace our basic structure of agency and responsibility because scholarly activity itself relies on the use of that structure, just as much as all our other kinds of activity do. The practice of science is itself a practice. It has to be understood as a way of acting freely, deliberately, and responsibly. Otherwise it cannot be understood at all. Our speech is not alone in having to follow the vulgar, everyday pattern here. The central structure of our thought does so as well.

It is interesting how thoroughly the secondary, scholarly ways of thinking can vary from age to age. Berkeley disowned the vulgar from a position very different from today's. Berkeley was an idealist who thought that it was the language of spirit that had won the battle. To support its triumph, Berkeley gave good reasons why the Materialist position couldn't be right, reasons why you cannot actually have a world of objects without subjects. Hume followed him by showing that it is actually quite hard to prove that the material world exists at all, if you once start to doubt it, and that in many ways it may be more economical to do without it.[14] In philosophy, in fact, the Idealist army won some well-deserved successes up to the end of the nineteenth century.

In the wider world, however, the fortunes of war went against it. Materialist troops seized the disputed territory and occupied it for most of the twentieth century, making assumptions are still the official creed recited in what T. H. Huxley called the Church Scientific. Yet, the reason why we are worrying about consciousness today is that this occupation hasn't worked very well either. The occupying troops themselves—the scholars working in areas where questions about conscious subjects keep coming up—are not happy. They report that the natives are restless. To drop the metaphor—the simplifications that materialism introduced, and that often conferred great advantages at first, have in many areas been exploited to their limit. Their attendant disadvantages are now coming forward. The approach must somehow be changed.

Having it both ways

How do we change it? A favourite suggestion for this at present is to bring out yet a third stuff, called Information, making it the successor to both Mind and

Matter. There are two reasons why this won't work. First, it looks plausible only because the word *information* is so wildly ambiguous that, when it's sloppily used, it can look like a stand-in for both its predecessors.[15] And second, it misses the point that this whole rivalry was misconceived in the first place.

There is no competition. Materialism and Idealism are not the names of warring tribes but the names of half-truths, neither of which can really be swelled into a comprehensive system. The real question is not 'what stuff is the world made of?' but, as Tom Nagel says, 'how to combine the perspective of a particular person inside the world with an objective view of the same world, the person and his viewpoint included'.[16] How can we bring these together so as to give a reasonably intelligible whole instead of leaving them loose so that they slide around and bang into one another?

It was natural to hope, of course, that a single conceptual scheme would provide a comprehensive, all-purpose explanation. This would have satisfied the demand for simplicity, which is one central aspect of our reasoning, and which many people still seem to see as its central guide. Seventeenth-century thinkers were convinced that, at the deepest level, the world was, as a matter of fact, simple in this way—a faith that seems still to have appealed to Einstein. And, of course, they followed that hope to tremendous effect. There could be no guarantee, however, that their conviction was right.

Today, many physicists see much less reason to make this wholesale assumption at all. For all we know, at the deepest level the world may go on for ever being more and more complex. The great reason for simplifying it is not that we have any advance guarantee of results. It is heuristic. It is just that, when simplifying does work, it greatly advances our understanding. It is therefore always worth while looking for simplicity. But that is very different from being sure that it is there to be found.

Many people today still believe that rationality offers us only the two competing schemes of dogmatic idealism and dogmatic materialism. They assume, not just that the universe has to have a single ultimate structure, but that that structure must be one of these two that we have already started to use. When we consider how schemes of thought have changed in the past and how much effort still goes into shaping them this does seem an extraordinarily bold belief. Rationality does not always demand simplification, and it certainly does not always demand elegance. Rationality is also concerned with finding a sane balance, with describing the world realistically and with adjusting means to ends. In science, and in enquiry generally, this adjustment means finding a conceptual scheme that works, one that suits your subject matter. Occam's razor cannot be the only tool on our bench. As Aristotle rightly said, it is the mark of an educated man to look for just so much exactness in every enquiry as the topic that you are enquiring about admits of—not asking a literary critic for geometrical proof or vice versa.[17]

Category trouble

In general, no doubt, we all admit this need to recognize complexity. But there is one particular kind of complexity that is quite hard to fit into current notions of what is scientific. This is complexity that involves more than one conceptual scheme and, still more, complexity involving concepts in different categories. That is the kind of difficulty that I tried to indicate in my opening example of the different kinds of map. Somebody who tries to account for the lines on the political map in physiographic terms will not succeed, and will not get on any better if they try to reduce physiographic language to something more basic still, such as particle physics. Very simple maps are certainly available and they do have their attractions, as was pointed out in *The Hunting of the Snark*:

> 'What's the use of Mercator's North poles and Equators,
>   Tropics, Zones, and Meridian Lines?'
> So the bellman would cry: and the crew would reply
>   'They are merely conventional signs!'
>
> 'Other maps are such shapes, with their islands and capes!
>   But we've got our brave Captain to thank'
> (So the crew would protest) 'that he's bought us the best—
>   A perfect and absolute blank!'[18]

But the uses of such maps are limited. For instance, if we want to know the explanation of that mysterious straight line on the political map, which in itself appears so simple, the only place where we can find it is in the history of a certain treaty and in the colonial system that lies behind that treaty. Here, explanation has to move outwards to that much wider and more complex background, not inward to the atomic structure. Treaties cannot be explained in terms of contours or vegetation, or electrons, nor of the neurones of the people who make them, any more than vice versa. They are not just a kind of casual folk shorthand for those things either. What is radically needed at this point is systematic talk about human history and human purposes. And this is talk that proceeds from a quite different angle.

This discontinuity between different viewpoints and different languages is not something imposed by philosophers. It is not something alien brought into science from the humanities. It is an unavoidable trouble that is well known to arise often when people from several different disciplines have to work together on a common problem. Modern specialization has greatly increased it by multiplying and narrowing the disciplines. It can arise, too, within a single enquiry, about the ambiguity of some abstract term such as time, space, evolution, infinity, particle, wave, or function. Such terms have various possible meanings, arising from different contexts of use, meanings that have to be disentangled when their role in a particular science begins to cause trouble.

Our difficulties about consciousness are simply more extreme examples of category trouble than most of these. This gap between the concepts that are used mainly from the subjective viewpoint—including those that deal with action—and those used from the objective one is indeed probably the most puzzling conceptual gear-change that we ever have to deal with. But even the minor clashes can be quite awkward, because their conceptual background always needs attention as well as their factual foreground. And sorting out conceptual tangles cannot help being philosophical business.

That does not mean, of course, that it can be done only by philosophers. Scientists often do it very well. In fact, the greatness of great scientists has often centred on just this skill in clarifying crucial concepts, especially in resolving category-problems that arise about them. Until lately, these thoughtful scientists were quite content to say that they were doing philosophy. They considered this as one of the things that they normally had to do. They knew when they were using philosophical tools. People such as Faraday and Maxwell, Bohr and Einstein, Heisenberg, Darwin, T. H. Huxley, and J. B. S. Haldane understood that every branch of science has its own philosophical problems They knew that these sciences very often need help from methods other than their own in resolving them, and they readily acknowledged that outside help.

The conclusion surely is that problems are not private property. They belong to anybody who can help to solve them. The advance of specialisation makes it harder to grasp this today, but it certainly does not make it less necessary. The fact that conscious beings are found in the world is a natural fact like any other. We have to accommodate it somehow. While we ignore it, we are not just left with a divided world but also with a disastrously divided notion of ourselves.

Notes and references

1. See, for instance, Jeffrey Gray's somewhat indignant assertion of a territorial claim for science against philosophy in *Journal of Consciousness Studies*, **2**, 8 (1995).

2. This is a main theme of Skinner's last book *Beyond freedom and dignity* (Penguin, Harmondsworth, 1973).
3. The systematic doubt by which he reached this conclusion is beautifully set out in brief in his *Discourse on Method*, Parts 4 and 5 and more fully in his *Meditations*.
4. See Jean-Paul Sartre, *Existentialism and humanism* (trans. Philip Mairet; Methuen, London, 1948).
5. See the end section of *The anthropic cosmological principle* by John D. Barrow and Frank R. Tipler (Oxford University Press, Oxford, 1986). I have discussed the worldview that lies behind these bizarre proposals in *Science as salvation* (Routledge, London, 1992), pp. 19–29 and 195–218, and in *Utopias, dolphins and computers* (Routledge, London, 1996), ch. 12, 'Artificial intelligence and creativity'.
6. 'The Kingdom of God' by Francis Thompson.
7. My first book, *Beast and man* (new edition Routledge, London, 1995), was largely concerned with the need to put ourselves together again by mending the Cartesian gap. It contains several discussions of Sartre's separatism (see under his name in the index). I have returned to the whole topic in *The ethical primate* (Routledge, London, 1995) and in an article called 'One world, but a big one' in *Journal of Consciousness Studies*, **3**, 500–15 (1996).
8. A. E. Housman, 'A Shropshire Lad', xxxii.
9. See I. Eibl-Eibesfeldt, *Love and hate* (Methuen, London, 1971), pp. 11–13 and 208–16.
10. David Hume, *Treatise of human nature*, Book 1, Part 4, section vii.
11. Rudyard Kipling, 'The two-sided man'.
12. See, for instance, *Beyond freedom and dignity* (note 2), pp. 102–3 and 111–12.
13. George Berkeley, *Principles of human knowledge*, section 51; see also sections 37 and 38.
14. Hume, *Treatise on Human Nature*, Book 1, Part 4, section ii.
15. A point well brought out by Raymond Tallis in a sharp little book called *Psycho-electronics* (Ferrington, London, 1994).
16. Thomas Nagel, *The view from nowhere* (1986, Oxford University Press), p. 3.
17. Aristotle, *Nicomachean Ethics*, book I chapter 3.
18. Lewis Carroll, *The Hunting of the Snark*, Fit the Second, 'The Bellman's Speech'.

CHAPTER 9

Towards a theology of consciousness

FRASER WATTS

In this chapter, I will try to outline a theological approach to mind and consciousness. It will certainly not be an approach that sets its face against the emerging scientific and philosophical understanding of consciousness represented in this volume. Rather it will seek to build on that growing understanding, reflecting on it in the light of theological concerns, and extending and recasting it from a theological perspective. This enterprise reflects some basic, general assumptions about how to do theology. The theological task is one of examining and reformulating traditional religious understandings in the developing contemporary context. Much of theology is thus interdisciplinary. Other disciplines contribute much of the context in which contemporary theology must be done. My particular concern is doing theology in the context of science.

Theology brings its own insights and concerns to the dialogue with science, but it must also listen to what science has to say before it can develop a constructive discourse. This chapter is organized about four theological concerns:

1. The question of how far God can be thought of as a centre of consciousness.

2. The nature of theological and mental discourses and how such discourses relate, respectively, to natural and material discourses.

3. The nature of 'soul', and what can be learned about the nature of soul from the parallel issue of the nature of mind.

4. The role of consciousness in religious life and, in particular, what we mean by 'religious' experience

The first theological insight about consciousness to be considered is thus about God. It is often thought that he can be understood in some sense as a mind, rather like a human mind, or as a centre of consciousness. The problem with this analogy is that God is not embodied like human beings, so the analogy is problematic from the outset unless we make the 'dualist' assumption that mind is, or can be, independent of body. Thinking about God as mind has thus led some theologians to believe that they need to subscribe to dualist views of mind and consciousness. I will suggest that theology need not take such a dualist turn.

In the second part of this chapter I examine how discourses about the human mind and brain relate to each other. I will argue that neither is dispensable, but that they represent complementary perspectives. That viewpoint provides a helpful model for relating theological and natural discourses to one another. In the dialogue between theology and science something useful can be learnt from philosophical psychology about how brain and mind perspectives relate to one another.

In the third section of the chapter I consider the implications of discourses about brain and mind for how we should think about soul. Soul is too often thought of as a thing. but if consciousness is an emergent property of our physical brains, perhaps soul should be thought of likewise. I will suggest that to talk about our 'souls' is to talk about ourselves from a particular perspective, not about something distinct from the rest of us.

In the final section I consider the religious significance of consciousness. It is through consciousness that human beings become aware of God; consciousness has a pivotal place in the relationship between God and humanity. This leads to a concern with the particular aspects of consciousness that are important from the religious perspective. There are, however, acute difficulties in deciding what we should mean by religious experience, and what is distinctive about religious consciousness.

The analogy between God and mind

In examining the analogy between the mind of God and the human mind, it is helpful to begin by looking at an argument that has been advanced periodically—namely that the existence of human consciousness provides an argument for the existence of God.

Some theologians, such as Swinburne (1979), have started from the assumption that it is impossible to give a satisfactory account of how consciousness emerges from matter and brain. That leaves a puzzle as to how consciousness arises. Swinburne then seeks to resolve the puzzle by introducing God as the source and origin of human consciousness. If the assumption is accepted that God is the source of the human mind, the existence of the human mind can then in turn be used as the premise of a further argument for the existence of God. By that stage, of course, there is a serious risk of circularity.

This argument has a long lineage. For example, in the seventeenth century, John Locke (1690/1960) in his *Essay concerning human understanding* argued that matter by itself could never produce thought, any more than it could produce motion. Thought must therefore have come from an eternal source, which must necessarily be a 'cogitative' being, who turns out to be God.

This is an approach to the theology of consciousness that I find misconceived and unconvincing. It may be helpful in clarifying what I am *not* saying in this chapter to explain why I do not follow it. I want it to be clear that the theology of consciousness need not embrace philosophical dualism. I am not necessarily seeking to refute dualism, but I am urging that God can be conceptualized without making dualist assumptions. One of my objectives in this chapter is thus to indicate the lines along which a non-dualist theology of consciousness can be formulated.

My first objection to Swinburne's argument is that, though it is true that we do not yet have a fully specified account of how consciousness might have evolved from brains of increasing complexity, there is no justification for concluding that such an account could not be offered. Any account of the neural basis of consciousness has to remain somewhat speculative until it has been elaborated more fully, and supporting evidence adduced. But some kind of emergence view of the origin of consciousness seems to me, as to most other people the one most likely to be correct. I certainly would not want to build a theological position on the premise that material theories of consciousness had failed.

It is worth emphasizing that, with each passing year, it becomes less appropriate to say that we have no idea how consciousness might arise from the brain. Admittedly, we do not have a complete account, and we certainly do not yet have one that we have sufficient reason to accept as correct, but there are several candidates in the field—such as, for example, the ideas of Edelmann (1992) and Crick (1994), and those summarized in this volume. We can no longer say that we have no idea how the brain might give rise to consciousness. Incidentally, I take Penrose's (1989) theory of distinctiveness non-computable physical processes in the brains as the basis of consciousness as a specific kind of emergence theory. He contrasts his position with emergence theories that see the complexity of the brain as the critical requirement for consciousness, and proposes a different kind of quantum basis for consciousness. To me, however, that seems a disagreement about exactly what is required for consciousness to emerge, rather than a disagreement about whether or not consciousness has some kind of physical basis. Penrose, I believe, is arguing for the non-algorithmic nature of consciousness, rather than its non-material basis.

In classical theism, care is taken not to regard God as one being among many, distinct from other beings, but comparable to them. I suggest that care should also be taken not to ask which being is responsible for a particular event, God *or* some human being. My general position, which I will elaborate in the next section of this chapter, is that theological and natural accounts should be seen as complementary rather than as alternatives. Applying this general principle, I submit that it is a mistake to ask whether we should attribute a phenomenon such as consciousness to God *or* to nature. The relevance of God to consciousness thus does not depend on the failure of natural explanations of consciousness.

Although philosophical dualism has gone out of fashion almost everywhere, it retains particular attractions for some theologians. Swinburne's argument for the existence of God on the basis of the assumption that there is no adequate natural explanation of consciousness is an example of this. It involves, I suggest, an over-literal view of the analogy between the mind of God and the human mind. Nevertheless, there is a close parallel between the issues about mind faced in philosophical psychology and the issues about the nature of God faced in philosophical theology.

The analogy between the two sets of issues is well brought out by Taliaferro (1994) in his *Consciousness and the mind of God*. In dualist theology, the link between the mind of God and the human mind is taken both ways. If a dualistic view of the human mind has been accepted, then it is easier to conceive God as a kind of disembodied mind too. Equally, if the idea of disembodied mind has been accepted in the case of God, then it is a modest further step to accept substance dualism in the case of human beings. Further, a dualistic view of mind may be attractive precisely because it links human beings to the mind of God. If the human mind is a substance relatively independent of matter and additional to it, that may seem to make human beings seem more similar to God, more open to him, more capable of becoming like him, than if an account of human beings is given in purely natural terms.

All talk about God is analogical of course, and the analogies are always rather inexact. So let us now reconsider the adequacy of mind as an analogue for God in the light of contemporary scientific and philosophical understandings of mind. Up to a point it is a helpful and appropriate analogy, but it has serious limitations if pressed too far.

The human mind is selective. Attention is by definition selective; we attend to some things at the expense of others. Memory is also selective: at any one time we do not consciously remember all the things that we are capable of remembering—specific memories are reconstructed as they are required. One of the implications of speaking of God as omniscient is that this kind of selectivity is not a feature of the divine mind. He knows all, attends to all, remembers all. In this the mind of God is so fundamentally unlike a human mind that the analogy comes under severe strain. God's mind can not be likened to a human mind, but with enhanced power: an omniscient mind would be organized on such radically different principles from a human mind that it would be nothing like ours.

Some of the classical attributes of God, such as omnipotence and omniscience, seem designed, in part, to be warning signs about the inadequacy of the analogy between the human mind and the mind of God. Human beings know and act but they are not omnipotent and omniscient. A good deal of what may appear to be the classic description of God, however, should be seen rather as a warning about how *not* to think and talk about God. As Lindbeck (1984) and others have argued, many theological statements are best assumed to be grammatical, even when they appear to be descriptive. They can be seen as regulative—setting boundaries to the appropriate talk about God, and warning against inappropriate, idolatrous talk.

Those who have thought about God as a mind have probably been thinking of the mind as a seat of consciousness, as a focal point of awareness. This corresponds to seeing the human mind also as a centre of consciousness or, in Ryle's (1949) famous phrase, as a 'ghost in the machine'. Implicit in this view is the image of sensory processes eventually reaching the 'little man' at the centre, who then makes decisions and sets in train motor actions. I do not need to rehearse the philosophical problems associated with this view. It is hard to see how to formulate the causal connections required to establish consciousness as a link in an otherwise-physical causal chain. What I want to emphasize is that this is not the only way of thinking about consciousness. Consciousness can be seen, not as a homunculus, but as an emergent property of a distributed system.

So far, however, there has been very little theoretical work on how consciousness might emerge from a distributed cognitive architecture. There are two reasons for this. One is that the dominant scientific interest is currently in how the brain produces consciousness. This is a perfectly legitimate question but my hunch is that it might be more fruitful for the time being to concentrate on the intermediate question of how a cognitive architecture might give rise to consciousness. The other important reason for the neglect is that, until quite recently, the dominance of the mind–computer analogy in psychological theorizing distracted attention from the theoretical problem of consciousness. Because computers are not conscious, psychological theorizing in the tradition of classical artificial intelligence has had a vested interest in assuming that consciousness is nothing more than an unimportant epiphenomenon.

That is now changing, and there is growing interest in how consciousness might arise within the cognitive architecture. Teasdale and Barnard (1993) have set out one approach to cognitive architecture that embodies explicit assumptions about where and how consciousness arises within the system. More generally, the introduction of connectionist modelling in psychology (see Bechtel and Abrahamsen, 1991), while not in itself the holy grail, has at least given us a general approach that is better adapted to handling emergent properties of the mind.

This developing notion of human consciousness as an emergent property of a distributed system makes it quite unlike what theologians have traditionally wanted to say about God. Though there are pantheistic theologies that would see God as some kind emergent world mind, this is along way from classical theism. The move away from thinking about mind as a central seat of consciousness in human beings, first general and philosophical but now more specific and scientific, should thus bring caution to any tendency to think about God as a seat of consciousness like the human mind.

Another important development in thinking about the human mind, along with seeing consciousness as a property of a distributed system rather than as a homunculus, has been to emphasize how closely intertwined are passive and active mental processes. Homunculus-type conceptions always left a gap between awareness and action. The assumption seemed to be that all impressions and information had to be received by the 'ghost in the machine' before it decided how to respond. The reality is that active and passive processes are intertwined throughout, both neurologically and psychologically. There has been a move to embrace what has called a 'motor theory of the mind' (Weimer 1977). As Varela *et al.* (1991) have put it, the human mind is not only embodied, it is also 'enactive'.

Such lines of psychological theorizing have been moving to a point that is convergent with the insights of continental philosophers such as Gadamer, who would also see the mind as actively bringing forth meaning out of a background of understanding. Up to a point this is helpful theologically. Dualistic theology tends, as Lash (1984) has caustically put it, to see God as 'either an idea or a ghost'. He suggests that it is better to see God as a 'thinker' than as a thought. The modern emphasis on the enactive mind would accord well with this theological emphasis. It would also provide an analogy for how the omniscience and omnipotence of God might be closely intertwined, rather than being separate attributes. Action is inherently cognitive; knowing is enactive. This is how we currently think about the human mind. It also seems a theologically congenial way of thinking about the mind of God. By analogy, therefore, I suggest that God's knowledge of the world is not detachable from his sustaining activity within the world, and similarly that his activity in the world is intertwined with his knowledge of it. This seems to be one of the points at which the analogy between God and mind has become *more* fruitful in the light of contemporary, scientific concepts of mind. Nevertheless, most of the points I have made emphasize how distant and loose is the analogy between the human mind and the mind of God. It is not an analogy that can be pressed too far.

From this vantage point it is interestingly to look back historically at the rise of the analogy. As Craig (1987) brings out in *The mind of god and the works of man*, there was a period of about a 100 years during the enlightenment when philosophers were particularly preoccupied by the analogy between the human mind and the mind of God. We often lose sight of the fact that this was a relatively novel preoccupation of the early modern period, with the analogy between God and mind being taken much more literally and taken further than it had been in classical theism. The same period came to see God as a 'person' in a new way (Webb 1919), another analogy that was pushed much further than it had been previously.

It is an interesting matter for discussion in the history of ideas exactly why these analogies were so appealing at the beginning of the modern scientific era. Craig points out that it was a period that sought a path to absolutely secure knowledge, and laid great store by human rationality—and by objective 'spectorial' observation as the means to achieve that. It was perhaps in keeping with this intellectual mood that God should be conceived as a being who represented the ideal of completely secure knowledge, the ideal to which the human mind aspired.

With the wisdom of hindsight, however, each interlinked element in this early-modern intellectual enterprise has come to seem misconceived. It no longer sees tenable to think of mind as a substance that is wholly separate from the physical body. Equally, it does not seem tenable to see knowledge as wholly objective and 'spectorial', independent of the person who arrives at it. Equally, I suspect that theology needs to draw back from the theistic aspects of this intellectual project. Indeed, modern atheism may have arisen in part as a reaction to some relatively new, but scarcely tenable, ways of thinking about God (Buckley 1987). Among these is the idea that God is a mind, like the human mind.

Mind–brain discourse and theological–nature discourses

I suggested above that it was probably a mistake to ask whether human consciousness could be explained in natural terms *or* whether it required a different kind of explanation invoking God. This point leads on to a more general issue about the relationship between natural and theological accounts, which I consider in this section.

To set up natural and theological explanations as alternatives is, in the end, unsatisfactory for both. Unless we take the atheist course of eliminating theological accounts altogether, we are left with dividing phenomena into two classes. On the one hand, there are some phenomena to which theological explanations are relevant but natural ones are not. This is unsatisfactory from a scientific point of view because of the boundaries that are then set on the scope of scientific investigation. I believe that science has something of relevance to contribute to the explanation of everything, including religion, even though there will also be other non-scientific things to be said. The second class of phenomena, on this approach, are those to which natural explanations apply but theological ones do not. This gives theism a very unsatisfactory twist, reducing God to a 'now you see him, now you don't' kind of being, relevant to some things and events but not to others.

The alternative is to find a way of carrying natural and theological explanations side by side. Interestingly, contemporary philosophy of mind gives us a helpful parallel to this, for it shows how, in a similar way, mind and brain languages can be seen as complementary, not alternatives.

One of the topics in consciousness studies in which I have done scientific work is going to sleep, and it illustrates my point about parallel discourses very nicely. Going to sleep can be described in terms of the physical processes of the brain, usually in terms of how the electrical rhythms become slower and larger. Equally, going to sleep can be described in terms of the phenomenological changes that take place. Thought processes become more fragmentary and less deliberate, orientation is lost, and bizarre images may flit unwilled through the mind. Both descriptions are valid in their different domains, and they are complementary rather than alternative. It would make no sense to ask whether sleep represented a change in brain state *or* a change in mental processes. Manifestly, it is both, and it can described validly, but incompletely, in either discourse.

Incidentally, on the best available evidence, admittedly rather fragmentary (Foulkes 1966; Bosinelli 1991), there seems to be a rough, but only a rough, mapping of changes in consciousness on to changes in brain state as people go to sleep. Both sets of changes tend to follow an orderly sequence, but the correlation between them appears to be only approximate. You cannot infer exactly, from knowing where someone's brain changes have got to, what phenomenological stage they will have got to in going to sleep. The empirical evidence available so far makes it look unlikely that phenomenological accounts of going to sleep will ever be redundant.

This is a kind of empirical evidence that is relevant to the claim, made by eliminative materialists, that mind language is redundant and just a convenient shorthand way of describing what could be more exactly described in terms of brain language. This claim assumes a one-to-one mapping of brain states on to mind states. As far as going to sleep is concerned, it looks unlikely that there is such a one-to-one mapping. Of course, it is always possible to argue that if we had better and fuller descriptions of brain states, phenomenological descriptions would become redundant. The concept of a brain state is an odd one, however, in that we have no idea how to produce a full description of a brain state, especially not one that is independent of the mind state with which it is associated (Harré 1970).

I submit that the relationship between brain discourse and mind discourse provides a useful model for how natural and theological accounts can be seen as complementary rather than as alternatives (see Watts, 1998). Natural explanations of any phenomenon can be offered, including consciousness. Such accounts can even become complete in their own terms. Nevertheless, a theological account can, in principle, be offered alongside the natural account, complementing it and cohering with it in the way that mind accounts complement and cohere with brain accounts. Having an adequate natural account does not make a theological account redundant, any more than having an adequate neurological account of a change in consciousness makes a phenomenological account redundant.

Note here that atheism corresponds directly to eliminative materialism. Both seek to simplify a situation in which there are two complementary discourses by eliminating one of them. It is not just that one discourse is accorded primacy over the other. Rather, one discourse is held to be complete and exhaustive, and the other discourse held to be redundant. Both atheism and eliminative materialism seek to establish that their preferred discourse renders the other discourse redundant. The only real difference between atheism and eliminative materialism is that the latter is more programmatic. Materialists would admit that we do not yet know how to manage without mind discourse, whereas atheism holds that it is already perfectly possible to speak a completely natural, non-theological discourse. Even that difference may, however, be more apparent than real, in as far as natural discourses often contain more hidden theological presuppositions than is at first apparent.

An important reason for keeping both mind and brain accounts of changes in consciousness, such as going to sleep, lies in the relevance of mental and physical explanatory factors. Mind and brain act as a unity; it never makes sense to ask which is responsible for a particular outcome, with the implication that the other was not involved at all. We do not ask whether a successful student used his or her mind brain to pass an examination. We presume that both were involved. There is, however, a useful sense in which some explanatory factors for any particular event can be regarded as primarily physical and others as primarily mental. Revising for an examination is primarily a mental process, although brain changes are no doubt involved, but no one could bring about the desired changes except by the mental act of revising. Mary Midgley makes a similar point more fully in Chapter 8 of this volume.

There is an important set of arguments here, with supporting empirical evidence, that is not always given sufficient weight by eliminative materialists. Some actions and interventions occur primarily at the level of mind or action, and occur only at that level, and have demonstrable efficacy. One interesting example arises in the rehabilitation of neurological patients and concerns the benefit of rehearsing particular actions. Here again there are two discourses. The first is a discourse about mere bodily movement that could in principle be described in mechanical terms; the second is a discourse about human action, a discourse that is always implicitly intentional. (This is not an original point, it is a commonplace of the philosophy of human action.) The interesting fact is that it makes a great deal of difference whether rehabilitation patients seek to rehearse implicitly intentional actions or whether they merely rehearse the corresponding mechanical movements. As Marcel (1992) has pointed out, the former result in measurably better progress in rehabilitation.

I want to suggest that here, too, there is a helpful analogy with the relationship between natural and theological discourse. Although I have rejected the idea that some events fall wholly into the natural realm and others wholly into the theological realm, there may nevertheless be some events for which the theological discourse is primary. Just as revising for an examination needs to be seen primarily as a mental activity rather than as a brain exercise, and just as rehabilitation needs to be approached as the rehearsal of actions rather than of mere movements, so there may be some events that particularly need to be conceptualized in theological rather than natural terms.

This approach would yield, for example, a formulation of what might be meant by a 'miracle'. I would explicitly not want to suggest that natural explanations are irrelevant to miracles. In the past, there has been a tendency to see miracles as occasions when God overturns the laws of nature, though there has recently been a move away from seeing things in such terms (Polkinghorne 1989). Scientifically, it is difficult to see how there can be events for which natural processes fail to operate at all. Also, theologically, it is problematic—if one assumes that the laws of nature are God's laws—to see him overturning his own laws. This leads to an alternative approach to miracles that would see them as events in which the laws of nature operate in a special way, rather than being overturned.

One particular way in which this could be explicated, within the kind of two-discourse approach I have suggested here, is that miracles are events for which the theological discourse is primary in an unusual and special sense. A natural account of miracles could still be offered, but miracles could be seen as events for which such a natural account is more radically incomplete than usual and especially needs to be supplemented by a theological account. Miracles would thus be acts or events for which the theological account carrried descriptive and explanatory primacy over the natural one.

The relationship between brain and mind discourses as an analogue of the relationship between natural and theological discourses is one that could be pursued at greater length. I hope I have said enough to indicate the potential fruitfulness of the analogy.

'Soul'

I have already indicated that I assume that mind and consciousness have evolved within the natural world, and arise from the physical brain. I will now turn to the implications of this for how we should think about mind and, by analogy, how we should think about the soul.

Since the seventeenth century we have often fallen prey to unhelpful ways of thinking about mind. We tend to see the mind as a thing or a substance, separate from the brain but of comparable status. That is misleading. I am not, of course, denying the reality of our mental powers, but it is inappropriate to 'reify' the mind. Talking about *the* mind carries dangers; we are on safer ground in talking about mental phenomena and processes (the adjective is less misleading than the noun).

We also tend to think of the seat of consciousness as being a kind of man in the middle—what Gilbert Ryle dubbed a 'ghost in the machine'. The implicit picture is of all sensations and perceptions being conveyed to this central seat of consciousness, where 'we' decide what is going on and what we want to do about it. Having made our decisions, we set our actions in motion. This way of thinking results from a muddling together of mind-talk and brain-talk into a confused and misleading hybrid. It would lead us to try to pin down, for example, the exact whereabouts in the brain of this central seat of consciousness. In contrast, as I have already said, many scientists now assume that consciousness emerges from the brain as a whole, functioning as a 'distributed system', not from just one part of it.

Now, I suggest that we can make comparable points about soul to those that I have just made about mind. Some people have become overwedded to thinking about the soul as some kind of separate entity, almost a thing, that is mysteriously added onto the rest of our natural, material nature. Sometimes the soul has been seen as an immaterial entity created by God before a person's natural birth, but this has by no means been universally accepted. That way of thinking goes back to Plato. An alternative tradition, promoted by Aristotle, and very influential within the Christian heritage, sees the soul as the form and function of the body, rather than as a separate entity. The soul is the person described from a perspective other than the material one, rather than a separate and distinct entity.

The pressures to see the soul as a separate entity grew in the seventeenth century, and came from various sources. People wanted to describe God in increasingly specific terms and were attracted by the analogy between God and an immaterial human soul. Also, some people thought that emphasizing the separateness of the body and the soul, and the limited nature of the body, would strengthen the argument that we were more than our bodies. In the long run, that proved a very counterproductive argument. Finally, it seemed that our potential for immortality could be explicated in terms of the survival of the soul after death, though it is doubtful whether that is the classical Christian tradition.

I want to suggest, instead, that soul is not a separate entity but a qualitative aspect of the person. As the Jungian psychologist, James Hillman put it, 'by soul I mean a perspective rather than a substance' (1975). I am as wary of talking about *the* soul as I am of talking about *the* mind. Perhaps we can see soul as a kind of emergent property of our whole beings, an aspect of us, which, like our minds, is grounded in our natural being. If we were not natural beings with bodies and brains, we would not have come to have souls, any more than we would have come to have minds.

Yet there are some ways in which soul is not exactly parallel to mind. Consciousness is a universal property of human beings, apart from a few who suffer from serious forms of damage or disorder. It is an interesting question whether soul is universal in the same sense. I see strong reasons for wanting to say at least that every human being has a capacity for soul-life. This way of thinking has been an important foundation for the respect that we properly ought to have for each human being, and I would not lightly abandon it. There is, however, another point of view from which the soul-life of some people remains more of a potentiality than an actuality. There is a hollowness and superficiality about many of us; and where there is little depth, there is little soul. Let me guard against any suspicion that soul is here being identified with intellectual powers. On the contrary, it is striking how people who are intellectually unsophisticated can have remarkable depth of personality and qualities of soul. I argued in connection with mind that our mental life really does involve separate and distinct powers, and we could never capture what is important in our mental life by talking just about what was going on in our brains. That point would hold in an even more radical sense when it comes to talking about our soul-life. Human soul-life is an even more remarkably emancipated flowering of our natural nature than our mental capacity for consciousness.

Thus there seem to be important 'both/and' things to be said about soul. On the one hand, I have suggested, our soul-life arises from our natural nature; on the other, it reflects our openness to God, our potential for becoming more spiritual beings, for taking on something of the likeness of God. Because of this, it carries our hopes of eternal life. There is something Janus-like about soul. While still being grounded in the natural, it is pointing beyond the merely natural.

One of the questions to which this leads is how far our soul-life can develop properties that become relatively independent of the natural existence out of which it has emerged—whether soul can take on a degree of independence from our natural bodies. I have said that when new properties emerge, new laws usually emerge too. Higher-level properties are not usually entirely predictable in their operation from lower-level laws. This means that, even though soul has come into being from a natural basis, once it has come into being it may to some extent operate in radically new ways that are not fully explicable in material terms. It may thus come to radically transcend our natural nature.

Religious experience

The religious life can be seen as a kind of 'schooling' of consciousness, and I will turn now to some of the issues that this raises. I will confine myself largely to the Christian tradition, though there would be interestingly different things to be said about the role of consciousness in other religious traditions, especially about Buddhism (see, for example, Varela *et al.* 1991).

Christians often talk about having an experience of the presence of God and this is at the heart of what has traditionally been meant by 'religious experience'. There has been a good deal of survey research on religious experience, such as that of Hay (1987). The relevant questions are often rather broadly framed, and talk about 'a presence or power, whether you call it God or not, which is different from your everyday self'. About 30 per cent of the population admit to having had such an experience.

Note at the outset, however, the oddness of the way the question is phrased. It is very different from how we would ask whether people had seen, say, a rainbow, or even whether they had a pain. With almost any other kind of question, it could more readily be assumed that people at least understood what they were being asked about. There would be no need to add a phrase such as 'call it God or not'. The oddness of the way the question is framed reflects the unique difficulty of talking about the experience of God. We seem unable to assume any consensus about what kind of experience is being referred to by the available language.

This leads on to a theoretical issue about the nature of religious experience that is currently quite hotly debated, and it can be framed in terms of just exactly what is 'religious' about a religious experience. There are alternative answers to that question. One is that religious experience is experience with a qualitatively distinct content, that is experience of something or someone quite different from other non-religious experience. The alternative answer is that religious experience is experience of ordinary things that are interpreted in a specifically religious way, even though they need not be interpreted like that. Is it the object of consciousness, or its interpretation, that makes an experience religious?

In recent years, important contributions have been made to the development of the second of these accounts, that religious experience consists of the religious interpretation of a wide range of what would otherwise be ordinary experiences. The point has been argued in slightly different ways by Katz (1978), Proudfoot (1986), and Lash (1988). Against this, Forman (1990, 1994) has argued persuasively for taking subjective accounts of the immediacy of religious experience more at their face value as experiences of a qualitatively distinct kind.

There is merit in both positions. I am certainly happy to acknowledge the power and relevance of processes of interpretation in religious experience. But I am doubtful whether religious experience can be seen as 'nothing but' the religious interpretation of a wide range of otherwise ordinary experiences. Some empirical points are relevant here. Accounts of experiences, which are in many ways prototypical religious experiences, are sometimes given by people who have either had little exposure to the relevant tradition of interpretation, or who would not want to make use of it. Hay (1987), for example, found that a significant proportion of atheists described experiences that appeared in most ways to be 'religious' experiences, even though they may have been reluctant to interpret them in these terms.

More philosophically, an account of religious experience as 'nothing but' the religious interpretation of ordinary experience seems to me to be too reductionist. I am not ready to dismiss the idea that there is, in religious experience, a distinct mode of experiencing the world, distinct mental content, and probably distinct objects of consciousness. I think, in fact, we are again dealing with complementary discourses, and that the two accounts sit alongside one another. The interpretative story about how religious experience arises can sit alongside the story of it being the experience of God. Arguing for the importance of the process of interpretation in religious experience does not show that religious experiences are not *of* anything distinctive.

Peacocke (1993), in *Theology for a scientific age*, has contributed an interesting discussion of a somewhat related issue. He focuses on the distinction that has been made by Swinburne (1987) and others between those religious experiences that are mediated by the senses and those that are not. Peacocke is wary of the notion that some religious experiences are not mediated by sensory experience or brain processes. First, he points out that, given the unity of mind and brain, the physical brain must be involved in *all* cosiness experience of God, mediated by sensory experience or not. Further, all religious experience will inevitably be derived from, and interpreted in the light of, our sensory experience of the public and natural worlds. These points must, I think, be correct, and it follows that there is no possibility of direct, 'unmediated' contact between the mind of God and the human mind in the sense of contact that somehow bypasses the physical brain and is uninfluenced by ordinary perceptual processes.

Nevertheless, it is still possible that there is something unusual in the way consciousness operates in religious experience. Some would see an altered state of consciousness as making a critical contribution to religious experience. Incidentally, this approach is compatible with either an emphasis on interpretation as to what makes an experience religious or an emphasis on the qualitatively distinct nature of such experience. A religious mode of consciousness might in principle facilitate either the awareness of specifically religious objects of consciousness, or it might facilitate the specifically religious interpretation of experience. There are, in fact, good reasons for recognizing that religious experience can be facilitated by particular states of consciousness. As I have argued elsewhere (Watts and Williams 1994), relaxed alertness, and the kind of 'broad-band' consciousness in which it is possible to be aware of a range of different things simultaneously, seems to be conducive to religious experience.

Traditional wisdom about religious experience often emphasizes that it depends to some extent on suspending habitual modes of discursive thought. Religious experience is perhaps more likely to arise in unusual modes of consciousness in which normal modes of interpretation are suspended. This point can also be accommodated by the two fundamentally different theories of religious experience I have outlined. If normal modes of interpretation are suspended, that may lead to a non-interpretative mode of consciousness (as Forman would argue). Alternatively (as Katz *et al.* would argue), it might lead to people learning a particular way of interpreting certain kinds of experience.

Let me now try to put this issue in long-term historical context. I want to suggest that there may have been important historical changes in the nature of consciousness generally, and that these changes have important implications for religious experience. Some of the issues about religious experience that arise in our own time may stem from the fact that our habitual modes of consciousness are somewhat different from those that obtained when the classical religious traditions such as Christianity became established. It seems likely that there have been long-term historical changes in consciousness, and that these are related to the kind of religious experience people have, or indeed to whether they have religious experience at all.

In particular, there seems to have been a gradual change in the extent to which the outside world has been felt to be distinct from an inner, subjective one (Watts and Williams 1994, chapter 7). Our present way of sensing and conceptualizing things, in which the inner world is seen as sharply distinct from an outer one, is apparently a contingent one. It is not how things have always been felt to be, nor how they have to be (Neumann 1954). This sense of separateness from an external world seems to have increased sharply around the seventeenth century, the same period in which, as I have already pointed out, there was a particularly strong emphasis on the analogy between the mind of God and the human mind. The story since then, though it involves some complex twists and turns, is roughly one of a gradually increasing sense of interiority.

It is hard to be sure whether some of the changes that have apparently taken place were merely literary or philosophical, or whether they were more basic phenomenological changes. I am inclined to think that they were the latter. Indeed, experience is so much shaped by how we *think*, that it is hard to see how changing ways of thinking could leave actual experience unaffected.

This increasing sense of separateness from the environment has been intertwined with linguistic changes. The linguistic trends appear to go back much further than the dawn of the modern period. Many linguistic terms seem to have begun by being 'double-aspect' terms, referring both to something in the external world and to an aspect of conscious experience. The tendency, however, has been for words to come to refer to something inner or something outer, but not to both (Watts and Williams 1994, chapter 9). This is of considerable theological interest in view of the extent to which religious experience often links the inner and the outer. A word such as 'light' continues to function in religious discourse as a 'double-aspect' term. Religious discourse uses double-aspect concepts, such as Charles Wesley's 'Christ whose glory fills the skies'. Indeed, double-aspect thinking and experience seem to be characteristic of religion. The erosion of double-aspect language, however, and the increasing sense of separateness of inner from outer, has made religious experience of this traditional kind, based on double-aspect thinking, less available to us in the modern world. The debate about the 'demythologization' of religious language associated with Bultman (1961) can be seen as arising from this demise of double-aspect thinking (see Watts and Wililams 1994, chapter 9).

We live, as I have said earlier, in the age of the 'enactive' mind. We have become aware of how active our minds are in constructing our experience. This has become clear from scientific research in cognitive psychology. The continental philosophical tradition of 'hermeneutics' is making a similar point. Yet another facet of this change is the tendency, to which Craig (1987, p. 229) has drawn attention, for 'practical concepts to invade areas previously thought of as purely theoretical', something that he calls the 'Practice Ideal'.

For those who value religious experience, the question is what forms of religious experience are available to a consciousness in which double-aspect thinking has been replaced by an enactive mind. In general terms, it will clearly be a kind of experience in which the work of interpretation plays a key place. I would suggest that it is a critical feature of the religious consciousness that it involves a sense of close linkage between the inner and outer. In the contemporary world, however, this link will probably have to be achieved in a way that is quite different from which obtained in old double-aspect thinking. For the enactive mind, such a link can probably only be achieved 'enactively', through processes of interpretation. So I return to endorsing, in a rather different way from Katz *et al.*, the role of interpretation in religious experience. As I see it, however, enactive, interpretation-based, religious consciousness is something of a possibility or an ideal, rather than a current reality.

I would not want to pit this way of looking at religious experience as an act of human interpretation against theological accounts that see it as a gift of God. Rather, I would want to hold together, as complementary discourses about religious experience, a theological discourse about it being the awareness of the presence of God, and a more natural discourse about it arising from the human work of interpretation. It makes no sense to me to choose between those two discourses about religious consciousness, any more than asking whether my students used their minds or their brains to pass their examinations.

Conclusion

Two broad themes have dominated this chapter. One is about the nature of complementary discourses. In the second section, I suggested that the relationship between mind and brain discourses is rather like that between theological and natural discourses. I argued against the tendency to eliminate one of a pair of complementary discourses, as both eliminative materialism and atheism do in their different ways. I also argued against pressing too hard the question of whether a particular event or phenomenon is to be explained in terms of one discourse or the other. Thus, I submit, for example, that it is unhelpful to ask whether human consciousness is to be explained in natural or theological terms, or whether religious experience should be seen as arising from a religious mode or interpretation as the experience of a qualitatively distinct domain. There is nothing inherently incompatible in these pairs of approaches.

The second broad theme was about the nature of mind, especially about how the human mind is active in shaping experience. Although I expressed considerable doubt about the value of the analogy between the human mind and the mind of God, this is one point at which it holds quite well. In as far as God is like a mind at all, he is an enactive mind. This theme recurred in the final section where I suggested that religious experience, if it was to occur at all in the context of modern consciousness, would have to arise from an active religious interpretation of experience. But I see this more as an aspiration than a current reality, and certainly not as showing that religious experience has no qualitatively distinct object or content.

References

Bechtel, W. and Abrahamsen, A. (1991). *Connectionism and the mind.* Blackwell, Oxford.

Bosinelli, M. (1991). Sleep-onset update. In *The mind in sleep: psychology and psychophysiology* (ed. S. J. Ellman and J. S. Antrobus), pp. 137–42. John Wiley, New York.

Buckley, M. (1987). *At the origins of modern atheism.* Yale University Press, New Haven.

Bultman, R. (1961). New testament and mythology. In *Kerygma and mythos* (ed. H. Bartsch). Harper, New York.

Craig, E. (1987). *The mind of God and the works of man.* Clarendon Press, Oxford.

Crick, F. (1994). *The astonishing hypothesis: the scientific search for the soul.* Simon & Schuster, London.

Edelmann, G. M. (1992). *Bright air, brilliant fire.* Basic Books, New York.

Forman, R. K. C. (ed.) (1990). *The problem of pure consciousness.* Oxford University Press, New York.

Forman, R. K. C. (1994). Of capsules and carts: mysticism, language, and the Via Negativa. *Journal of Consciousness Studies,* 1, 38–49.

Foulkes, D. (1966). *The psychology of sleep.* Scribners, New York.

Harré, R. (1970). *The principles of scientific thinking.* Macmillan, London.

Hay, D. (1987). *Exploring inner space* (2nd edn). Mowbray, Oxford.

Hillman, J. (1975). *Re-visioning psychology.* Harper and Row, New York.

Katz, S. T. (ed.) (1978). *Mysticism and philosophical analysis.* Oxford University Press, New York.

Lash, N. L. A. (1984). Materialism. In *A new dictionary of Christian theology* (ed. A. Richardson and J. Bowden), pp. 353–4. SCM Press, London.

Lash, N. L. A. (1988). *Easter in ordinary.* SCM Press, London.

Lindbeck, G. A. (1984). *The nature of doctrine.* SCM Press, London.

Locke, J. (1690/1960). *An essay concerning human understanding.* Collins, London.

Marcel, A. J. (1992). The personal level in cognitive rehabilitation. In *Neurological rehabilitation* (ed. N. von Steinbuchel, D. von Cramon, and E. Poppel), pp. 155–68. Springer-Verlag, Berlin.

Neumann, E. (1954). *The origins and history of consciousness.* Routledge and Kegan Paul, London.

Peacock, A. (1993). *Theology for a scientific age.* SCM Press, London.

Penrose, R. (1989). *The emperor's new mind.* Oxford University Press, Oxford.

Polkinghorne, J. C. (1989). *Science and providence.* SPCK, London.

Proudfoot, W. (1986). *Religious experience.* University of California Press, Berkeley.

Ryle, G. (1949). *The concept of mind.* Hutchinson, London.

Swinburne, R. (1979). *The existence of God.* Clarendon Press, Oxford.

Taliaferro, C. (1994). *Consciousness and the mind of God.* Cambridge University Press, Cambridge.

Teasdale, J. D. and Barnard, P. J. (1993). *Affect, cognition and change: remodelling depressive thought.* Lawrence Erlbaum, Hove.

Varela, F. J., Thompson, E., and Rosch, E. (1991). *The embodied mind.* MIT Press, Cambridge, Mass.

Watts, F. (1998). Science and religion as complementary perspectives. In *Rethinking theology and science: six models for the current dialogue* (ed. N. H. Gregersen and J. W. van Huyssteen). Fortress Press, Minneapolis.

Watts, F. and Williams, M. (1994). *The psychology of religious knowing.* G. Chapman, London. (Previously published 1998, Cambridge University Press.)

Webb, C. C. J. (1919). *God and personality.* Aberdeen University Press, Aberdeen.

Weimer, W. B. (1977) A conceptual framework for cognitive psychology; motor theories of mind. In *Perceiving, acting and knowing: towards an ecological psychology* (ed. R. Shaw and J. Bransford), pp. 267–311. Hillsdale, New Jersey.

CHAPTER 10

Recovering contingency

NICHOLAS LASH

At the end of *The rediscovery of the mind*, John Searle offered some 'rough guidelines' towards that goal, of which the 'fourth and final' one was that 'we need to rediscover the social character of the mind'.[1] He had already noted, at the midpoint of his argument, that there are two subjects 'crucial to consciousness', about which he had little to say because he 'did not yet understand them well enough': 'temporality' and 'society'.[2] If we take the 'modern' world to mean the world constructed in Western Europe and North America in the seventeenth and eighteen centuries then, notwithstanding the bewildering imprecision with which the epithet 'post-modern' is at present strewn around, it seems to me that, in identifying those two subjects as 'crucial to *consciousness*', Searle showed himself to be, if not post-modern, then at least a most *un*modern kind of thinker.

Much of this chapter will amount to little more than grace-notes to his argument; hints and indications as to why I think he is quite right to point in these directions and as to the effect that moving in them might have on our accounts of who and what we are, and of how we are the kinds of things we take ourselves to be. Before getting under way, however, I have two anecdotes, one general remark, a morsel of polemic, and an explanation.

Did society come first?

Many years ago Lord Longford told an uncle of mine that, in his youth, while still a member of the Conservative party, he was invited to lunch by Stanley Baldwin. As they strolled in the grounds after their meal, the starry-eyed young politician asked the elder statesman: 'Tell me, sir, is there any one political thinker who has been especially influential in the development of your own thought?' After some reflection, Baldwin replied: 'Yes, there is; there is So-and-so. You see, what I learned from him, and I have never forgotten it, is that, where relations between society and the State are concerned, society comes first. Or was it the other way round?'[3]

It is my impression that, where relations between human individuals and the larger natural and social formations of which they form a part are concerned, much of our current scientific and philosophical thinking is in a disturbingly Baldwinesque condition. We know that we are fragile, interactive products of biology and circumstance, wholly dependent at every level of existence upon the complex webs of causes and effects of which we form a part, and yet, in our discussions about 'human identity', the focus of attention is usually an individual agent as apparently autonomous as any 'early modern man' might have supposed himself to be.

My second (and related) anecdote concerns Geoffrey Lampe, who was Ely Professor of Divinity in Cambridge from 1959 to 1971, and then Regius Professor until 1979. Enormously learned in the writings of the Greek and Latin Fathers, formal or 'grammatical' considerations were quite alien to him. We were close friends, but could never get within earshot of each other on the topic (to which I shall return) of 'subsistent relations' in Trinitarian theology. As his motto in this matter, we could take this sentence from his 1976 Bampton Lectures, *God as spirit*: 'If there are relations there must be entities that are related'.[4]

My general remark concerns what might properly be called the politics of our topic. Human identity is not simply a natural 'given' in the sense that the identity of a hydrogen atom would appear to be (where by 'given' I mean, at this point, no more than 'datum'; although whether, and in what sense, human identity is not merely Latin 'datum' but also English 'gift' is a question well worth serious philosophical and theological consideration). Human identity is not simply a natural 'given' because (however unwelcome this claim may be to spirit–matter dualists) the constituents of what we call 'culture' are thereby constituents of human 'nature', and hence of our identity,[5] and culture—the ways we live together, the symbol-systems and the cities that we fashion, the networks of information and control that we construct—is as much a project as it is a given fact.

I say that this remark concerns the politics of our topic because the specification of the project that human being is is not the prerogative of any particular group of human beings. Hydrogen does not decide what being hydrogen will be. In contrast, determining what human being will be is—however severe the limits within which, in fact, we operate—part of what it means to be a human being. It follows, as the Indian theologian Felix Wilfred remarked recently, that 'Defining the human is not and cannot be the prerogative of one civilization or one people'.[6] (My hope is that this remark may flush out the residual Cartesianism in some of my scientific friends who suppose themselves quite free from the disease!)

Next, a morsel of polemic. As a Catholic theologian with quite conventional views on Christian doctrine, I welcome John Searle's 'biological naturalism' as wholeheartedly as I do Gerald Edelman's attempts to put the mind back into nature.[7] Which may surprise Searle, who curiously supposes that people who believe as I do are likely to be Cartesians suffering from the 'antiscientism that went with traditional dualism, the belief in the immortality of the soul, spiritualism, and so on'.[8] In blissful ignorance, it seems, of the extent to which popular substance–dualist construals of the relationship between the human body and its life, or soul, had for centuries been an embarrassment to mainstream non-dualist Aristoteleanism (to say nothing of being incompatible with the Jewish anthropology that the Christian Scriptures breathe), and hence of the extent to which Cartesian dualism is a recent disruptive *innovation* in the history of Christian anthropology, he blithely classifies Cartesianism amongst the 'traditional religious conception[s] of the mind'.[9]

Contrast this muddle with the more plausibly traditional conception of the mind that Anthony Kenny once offered in order to indicate 'the magnitude of the Cartesian revolution in philosophy', and to show how dramatically the place where Descartes drew the boundaries of the mind differed from 'where they had been drawn by his predecessors in antiquity and in the Middle Ages, in the tradition going back to Aristotle'. Kenny defined the mind as 'the capacity for behaviour of the complicated and symbolic kinds which constitute the linguistic, social, moral, economic, scientific, cultural, and other characteristic activities of human beings in society'.[10]

Searle's insouciant misdescription of the way that things have stood for most of Western cultural history is, it seems, due to the fact that he is one of those people who, in their 'deepest reflections', simply cannot 'take ... seriously' the opinions of people who believe in God; which fact about himself he is modest enough to characterize, at one point, as 'insensitivity'.[11] Here ends the first morsel of polemic.

Finally, to end these scattered introductory remarks, a word of explanation as to why I contributed to the conference of which this volume is a record. I am a professional theologian and amateur philosopher with an interest in the history of ideas whose scientific education came to a shuddering halt at the age of seventeen, with the result that I am only able to peer, with fascination, over the shoulders of people such as Searle and Edelman to find out what is going on inside my brain.

With the preliminaries disposed of, there are, in the body of this chapter, three things that I would like to do. First, to offer some reflections on the way that God is 'one', and on the esoteric notion of 'subsistent relations' in divinity. This will enable me indirectly to comment on 'the social character of the mind'. In the second place, it is worth asking how the 'temporal' character of consciousness got lost from view? Here I shall turn for help to Francis Bacon. Finally, I shall try to indicate why what I have called, in the title of my chapter, 'recovering contingency', is crucial to all our efforts to make better sense of who and what, as human beings, we are.

The way that God is one

I hope that the fact that, in this section of my chapter, I am going to talk theology, will not unduly distress those of you who find the enterprise quite uncongenial. Think of what I have to say as serving the same kind of purpose as those illustrative fictions which philosophers construct when they find thinking about the real world too difficult.

After all, Ludwig Feuerbach was by no means the first to notice that our images of God reflect the forms of our experience. When the people of Israel spoke of God as shepherd or as king, they knew what they were doing (as one can see by noticing that, very often, what they were doing was contrasting God's ways and ours: unlike other judges, God judges justly; unlike other shepherds, God brings back the strayed, strengthens the weak, and so on). If, as Emile Durkheim argued, religion is 'the system of symbols by means of which society becomes conscious of itself',[12] then one way of finding out how human beings understand themselves—what they take to be their nature, identity and purpose—will be by attending to the ways in which they speak about whatever it is that they most cherish, venerate, revere.

It is not uncommon these days for people to discuss whether or not there is 'a God'. In such discussion, godness or divinity is being taken as a nature name, the label for some kind or category of which there may or may not be an instance or some instances. Such speech has little in common with the grammar of the classic forms of mainstream Western Christianity (to which, for present purposes, I confine my illustration). God, in this tradition, is not best spoken of as an individual with a nature of some kind, for the One that alone is to be worshipped is beyond all categories, creates and comprehends all kinds.

The conceptual and grammatical resources for giving expression to this recognition (derived from Jewish insistence on the holiness of God) and for sustaining the disciplined negation it requires, were drawn, in part, from Neoplatonism. Christian thinkers, uncomfortable with the priority that the Neoplatonists ascribed to the One, as to the Good, beyond both 'mind' and 'being', may have varied their vantage-point, now laying the emphasis on goodness as the goal and consummation of all things' being and desire, now (as in Aquinas's case) insisting resolutely on the priority of being, but the central thrust was to assert the *identity*, in God, of goodness, unity, and being.[13]

Part of what is at issue here, of course, is the insistence that all things come from one good principle and go to one good end. In other words, it is as characteristic of Christian cosmology as of Jewish to reject those dualisms for which the world's existence is undying warfare between the darkness and the light, between order and engulfing chaos.

Moreover, lest Neoplatonic metaphors of emanation be too literally construed, Aquinas's ascription of priority to being is at the service of a strategy that sets consideration of the ways in which God does *not* 'exist' at the top of the agenda.[14] Thus, while recognition of contingency may properly awaken wonder, anyone who supposes that they have God in their conceptual or imaginative sights, that they know what God is like (or would be, if there were a God) is not, according to Aquinas, thinking about God at all.

But there is more to it than this. Although it is easy to lose sight of the fact when reading his tersely formal texts, Aquinas is not meditating, abstractly, on divinity or 'godness' but more concretely considering what is to be said of One confessed as Father, Word, and Spirit.[15] Thus, for example, he introduces a discussion of whether God is to be thought of as 'supremely one' by quoting, with approval, a passage from Bernard of Clairvaux in which Bernard passes through eight different kinds of unity—those pertaining, for example, to collections and to organisms, to marriages and to societies—before declaring that 'amongst all the things that rightly are called "one", the summit or high point ("*arx*") is occupied by the oneness of the Trinity'.[16]

Rhetoric, yes, but not mere verbal gesturing. The history of Christian theology is the history of attempts to think of one confessed as Father, Word, and Spirit, without collapsing what is thought of into a group of individuals. (If, by the way, we keep in mind the Durkheimian rubric that religion is the system of symbols by means of which society becomes conscious of itself, then we could—impressionistically—interpret what I have just said as implying that, wittingly or unwittingly, the history of Christian theology is the history of attempts to think of given meaning and common life, of message and community, without, on the one hand, divorcing meaning from its ground or, on the other, separating doctrine from life, the ideal from the real.)

'Philip said to him, "Lord, show us the Father, and we shall be satisfied".' To which Jesus replies: ' "Have I been with you so long, and yet you do not know me, Philip? He who has seen me has seen the Father".'[17] Thus it is that the message heard and seen and touched as Jesus of Nazareth, a particular Jew, was said to be one and the same as the message's utterer, as the one called God, Creator, Father. One and the same what? Distinctions were drawn, vocabularies constructed, with varied success. God, it was said, is three 'persons' in one 'nature'. But, even in Augustine, it is (according to Henry Chadwick) 'unclear whether we are being told that the three Persons exist in relation to one another, or whether relation is integral to the notion of Person'.[18]

That last phrase gives us the clue that I want to follow up. Nearly 900 years separate Augustine from Aquinas, in whose hands the doctrine of subsistent relations reached its most thoroughly worked out expression as a set of protocols with the aid of which we may try to speak of a God who exists as, and only as, the relations that God is. And yet the neatest illustration that I know of how this apparently austere and abstract formal theorem might be 'cashed' may be found in some comments of Augustine's on the Fourth Gospel.

'The Son' (according to that Gospel) 'can do nothing of his own accord, but only what he sees the Father doing'.[19] But what would it mean to say that the Son 'sees' the Father's work? At this point, I want to push a little further a suggestion that Searle makes about conscious subjectivity. According to Searle, 'We cannot get at the reality of consciousness in the way that, using consciousness, we can get at the reality of other phenomena'. The reason for this is that, as he says, 'where conscious subjectivity is concerned, there is no distinction between the observation and the thing observed'.[20] True, but the assumption that, elsewhere in our thinking, the distinction between the 'observation' and the thing 'observed' is always, or even for the most part, the *kind* of distinction (implying the kind of 'distance') that there is between the eye and an object that we literally 'observe', is quite unwarranted.

In other words, when we say that we 'see' this, that, or the other, what we often mean is that we 'see the *point*' about it, that we understand it. Throughout the Fourth Gospel, 'seeing' is thus used as a metaphor for understanding. And it is this that gives Augustine the clue that he needs: 'The way in which the Son sees the Father is simply by being the Son. For him, being from the Father, that is being born of the Father, is not something different from seeing the Father.' Or, as he put it elsewhere: '*Videndo enim natus est, et nascendo videt*'; 'In seeing he is born, and in being born he sees'.[21]

God's very identity, God's existence, simply *is* 'seeing the point' about Himself, the world, and us. And that relation works both ways: as the seeing of the point, and as the point that is thus seen. And so it is that if, in seeing Jesus, we see the point about him, we have, in seeing it, seen God. Which is what Philip failed to see.

I apologize for dragging you through what may have been, for some, bizarre and unfamiliar material. But I hope that, now, you see the point! Following, once more, my Durkheimian rubric, the suggestion is that a society which thus described—in terms of pure relationship, pure 'donation', without remainder—not merely, we might say, the 'character' of God, but God's very being, God's identity, must surely have had a sense of human identity, of what it is to *be* a human being, profoundly different from one which (like our own) for the most part takes it for granted that, in Geoffrey Lampe's expression; 'if there are relations there must be entities that are related'.[22]

I would, moreover, urge that it is questions of ontology, and not just of ethics, that are at issue here. Of course we are talking about ethics as well, for I take it to be an implication of my earlier reminder that human nature is both fact and *project* that the ontology of human being entails some set of ethical proposals or presuppositions.

I earlier described as 'Baldwinesque' our tendency simultaneously to speak as if human individuals were mere epiphenomena—particles of flotsam, or fragments of some vast machine or organism—and agents bearing the burden of autonomy. It would, however, have been just as accurate to describe this inconsistency as 'Thatcherite', because the prophet of a rugged individualism who insisted that 'there is no such thing as society' was no less adamant in her belief that 'you can't buck the market'.

My hunch is that there is no way through these well-known aporia except through the recognition that the two subjects picked out by Searle as 'crucial to consciousness'—namely, temporality and society—are, both historically and conceptually, inextricably interrelated. That being so, and having indicated my conviction that the history of Christian thought may contain neglected resources that might help us in our rediscovery of 'the social character of the mind', I now want to move on and, with a little help from Francis Bacon, suggested when and why it was that the medieval privileging of the relational became, almost literally, unimaginable.

People are stories

According to Searle, 'one of the keys' to the original development of a still widespread world-view that will be baffled by his contention that 'consciousness is just an ordinary biological feature of the world' was 'the exclusion of consciousness from the subject matter of science by Descartes, Galileo, and others in the seventeenth century'.[23] I now want to suggest that it might be useful to put a slightly different spin on the developments that he has in mind by viewing them in terms of what I would call the neutralization of memory or (if you will pardon the horrid word) the denarratization of knowledge.

Commenting on Lessing's pronouncement that 'Accidental truths of history can never become the proof of a necessary truth of reason',[24] Karl

Barth once said, 'This sentence does not say ... what Fichte later said: "It is only the Metaphysical and on no account the Historical, which makes me blessed" '.[25] Nevertheless, that is the direction in which Lessing points. In other words, by 1777 (when Lessing wrote those words) the wedge between 'event' and 'truth' had been so firmly driven home that learned men had quite forgotten that explanations take the form of stories, told by someone, and not by nobody, in one place and not nowhere, in some manner that might have been otherwise.

In 1750, Denis Diderot published his Prospectus for that Bible of French Enlightenment, the *Encyclopédie*. There we see the whole map of human knowledges ('*connoissances* [sic!] *humaines*') laid out in three columns: memory (for history), reason (for philosophy), and imagination (for poetry). Not the least striking thing about this map, however, is that, with the exception of a slight adjustment in the location of what Diderot called 'revealed theology', the entire scheme has been lifted, lock, stock, and barrel, from Francis Bacon's *Advancement of learning*, which was published in 1605, before Descartes, or even Galileo, had made their mark. It is, therefore, to Bacon's version of the map that I now turn.[26]

'Any one', said Bacon, 'will easily perceive the justness of this division that recurs to the origin of our ideas... it is clearly manifest that history, poetry, and philosophy flow from the three distinct fountains of the mind, viz., the memory, the imagination, and the reason; without any possibility of increasing their number. For history and experience are one and the same thing; so are philosophy and the sciences'.[27] The two features of his map to which I would draw particular attention are, first, the trivial place occupied by imagination and, second, its comprehensive dissociation of memory from argument, experience from reason.

Under 'imagination', Bacon listed only 'poetry', of three kinds: 'narrative', 'dramatic', and 'parabolic'. With this exclusion of imagination from the territory of serious enquiry, interest is thereby abandoned in the cognitive significance of metaphor and story-telling, of parable, polysemy, and paradox. There is, that is to say, a striking lack of interest in the concrete forms communication takes between human beings in existing networks of relationship. Bacon's mapping of the mind thus eloquently illustrates Amos Funkenstein's account of the way in which, at this period, the fusion of the late medieval passion for plain speech and single meanings with the Renaissance rediscovery of Stoic 'nature'—a world seen as homogenous through and through, made of one kind of stuff and driven by one set of forces—gave birth to a new ideal for the working of the human mind: namely, 'a science that has an unequivocal language with which it speaks and uniform objects of which it speaks'.[28]

'For history and experience are one and the same thing; so are philosophy and the sciences', the former being the province of 'memory' and the latter that of 'reason'. History is broken down into 'civil history'—which covers 'memoirs' and 'antiquities', the history of the church, of prophecy, and of providence—and 'natural history', comprising the story of the heavens, of meteors, of the earth and sea, of monsters, and of the agricultural, manual, and mechanical arts. Philosophy, on the other hand, is reason's study of God, and man—the body and the soul, the arts of conversation, negotiation, and state policy—and of nature: considered either 'speculatively', in physics, or 'practically', in magic and mechanics. Perhaps the most striking thing about this set-up, in which, as I said earlier, memory is set apart from argument, experience from reason, is the reduction of history to data—raw material for the working of the mind.

Items of information 'first strike the sense', said Bacon, 'which is as it were the port or entrance of the understanding'.[29] He seems to be sitting in a kind of capsule, a Tardis perhaps, receiving, through different apertures, raw materials to be processed: reviewed, considered, classified. There is no sense of his being part of a story, or set of stories, which have shaped his world and made things to be the way they are or seem to be; no sense of his being caught up in conversation, part of some larger set of narratives that he must enact, interpret, endorse, or struggle to revise. In other words, as early as 1605 we find Bacon inhabiting a world that is already recognizably 'modern' in that—if not temporality, then at least narratability—a sense of whence and whither, of a story with a plot, is ceasing to be constitutive of human being and of the being of the world. The 'ugly, broad ditch' that Lessing could not cross—between what happens to have happened and what must be true—is, it seems to me, already opening up.[30]

There is no small irony in the recognition that Lessing's ditch should turn out to have been little more than a figment of the modern imagination, because the truths of reason are never quite as necessary as those who formulate them may suppose, and historical contingency may bear the meaning of the world and, perhaps, the truth of God. (The nineteenth century, of course, resolved the dilemma in the other direction: not by recovering contingency, but by what Karl Popper called 'historicism', the identification of what happened with what must be so.)

It has taken us too long to learn, again, that there is no neutral vantage-point, no 'nowhere in particular', from which truth may be discerned and the pattern of right action estimated. Narratives take time and, as we learn to find our way between experience and understanding, between what came before and what now lies ahead, it is as constituents of and contributors to a 'story-shaped world' that we proceed.[31] Such formal systems as we may construct, in philosophy or science, are shaped, determined, coloured, by the narrative soil from they spring.

Because we bear responsibility, individually and socially, for the stories that we tell, the narratives that we enact, the lives that we perform, there is what we might call an autobiographical component to every story that we propose as true. Biographies are stories of lives other than our own, whereas to make the story mine—to claim or to acknowledge that this is what *we* have been up to, what we are doing, where we ought to go—is to render its narration autobiographical. What, with these remarks, I have been trying to do, and I hope that the attempt does not seem too far-fetched, is to indicate connections between the temporal and self-involving character of truthful speech and what Searle calls the 'first person' character of 'the ontology of the mental'.[33]

When did the privileging of the relational become, almost literally, unimaginable? The answer seems to be in the late sixteenth and early seventeenth centuries, when new maps of knowledge not merely (as Searle observes) excluded consciousness from the subject-matter of science but, in their systematic dissociation of memory from argument, of narrative from truth, obscured from view the social, self-involving, tradition-grown or 'conversational' character of the human quest for truth.

Recovering contingency

According to Searle, 'the deepest motive for materialism ... is simply a terror of consciousness', and 'the deepest reason for the fear of consciousness is that consciousness has the essentially terrifying feature of subjectivity'.[34] But why should subjectivity be terrifying? In being frightened of it, of what are we afraid? The answer would seem to lie in what Richard Bernstein called 'Cartesian anxiety'. 'Descartes' search for a foundation', said Bernstein, 'is the quest for some fixed point ... [and] the specter that hovers in the background of the journey' undertaken in the *Meditations* is 'not just radical scepticism but the dread of madness and chaos where nothing is fixed'.[35] A dread that Nietzsche also knew.

Here is Stanley Cavell, pushing the anxiety a little further back, behind Descartes, and locating it in the sense of radical solitude, of the absence of reliable relations:

> As long as God exists, I am not alone. And couldn't the other suffer the fate of God? It strikes me that it was out of the terror of this possibility that Luther promoted the individual human voice in the religious life. I wish to understand

how the other now bears the weight of God, shows me that I am not alone in the universe.[36]

But why should we suppose that any other, any others, are strong enough, sufficiently reliable, to bear such weight? And so, suspecting any such supposition, any such large-scale trustfulness, to be unwarranted, anxiety is unassuaged.

Against this background, it is not surprising that the interests of modernity should have alternated between reliance on 'objects'—amenable to our control, neutral and unthreatening in their computability—and the retreat to romantic subjectivity, to weekend-worlds of private feeling—that is, of course, for those with the resources to indulge this alternation. (Notwithstanding the resolutely anti-Cartesian character of Searle's programme, it does seem to me that his use of the terms 'objective' and 'subjective' is still too strongly flavoured with subject–object dualism.[37])

Meanwhile, the energy that fuels the fear, that feeds the terror of our solitude, takes flesh as violence, seeks safety through control. Mary Midgley has often pointed out how even some quite serious scientific works, going all unbuttoned in their final chapter, indulge in fantasies of the control of things by mind, which display a most unscientific dread of death, of flesh, of our contingency.[38] For which reminder that scientists are sinful human beings, she has sometimes been most unfairly charged with being 'anti-science'. Not that the texts she mentions would have come as much of a surprise to Nietzsche, who understood the will to power.

It is time, I think, to go back to the beginning, to the early modern terror of contingency. 'Le silence éternel de ces espaces infinis m'effraie'.[39] Was it the sheer scale of those spaces or their silence that most terrified Pascal? Insofar as it was the latter, we are brought back to the fear of solitude as the ground of our disquiet. But solitude occurs in context, the context of vast empty spaces through which silent systems turn. We are not simply on our own, it seems, but inexorably, necessarily alone.

'There is', says Searle, 'a sense of panic that comes over a certain type of philosophical sensibility when it recognizes that the project of grounding intentionality and rationality on some pure foundation, on some set of necessary and indubitable truths, is mistaken in principle.'[40] It is as if contingency were more than we can take. However, according to Gordon Michalson, in his study of Lessing, 'contingency is nerve-wracking only for those who have a stake in necessity'.[41] In which case, the recovery of contingency would seem to be, in part, a matter of surrendering that stake.

In 1916, at the age of twenty-seven, Ludwig Wittgenstein, during a lull in the fighting on the Russian front, wrote in his notebook: 'The wonderful thing is that the world exists. That there is what there is.'[42] I think it is worth trying to read that note without the background noise created by commentaries on the *Tractatus*; trying to read it, in other words, simply as an expression of wonder at the world's contingency, without invoking the thoroughly misleading category (in this context, and many others) of the 'mystical'.

What is the difference between Wittgenstein's wonder and Pascal's fear? The question is worth asking, if only because no less an authority than G. H. von Wright spoke of a 'trenchant parallelism' between the writings of Pascal and those of Wittgenstein. Con Drury (who mentions this) admitted that there is something in it, but was nevertheless more inclined to emphasize the differences between the two. 'Drury', Wittgenstein once exhorted him, 'never allow yourself to become too familiar with holy things.'[43] The suggestion is that, unlike Pascal, Wittgenstein cannot be suspected of 'fideism', which Drury takes to be a way of avoiding difficulties by too familiar acquaintance with the holy.

Perhaps Wittgenstein's wonder, unlike Pascal's, did not tighten into terror because he had no 'stake in necessity'; did not seek to ground intentionality and rationality 'on some pure foundation'. This, if I follow him, is a good part of the reason why Searle invokes Wittgenstein's authority for his notion of the 'Background'—that sum of 'capacities, abilities, and general know-how that enable[s] our mental states to function'.[44] The important point, for my purpose, being the emphasis upon contingency: 'The Background does not have have to be the way it is.'[45]

I spoke earlier of that alternation of subjectivity and a realm of calculated 'objects' with which the modern world has seemed so often stuck. In a recent essay in phenomenology, Jean-Yves Lacoste recast this deadly dialectic in terms of the relations between the earlier and the later Heidegger.[46]

With the Heidegger of *Sein und Zeit*, we find ourselves 'thrown' into a 'world': a worldly world, a *'saeculum'*, a secularized or disenchanted place. This recognition that we are 'strangers' lacking direction and (as Paul said) 'without God in the world',[47] may be intermittent, but it is as old as Greece and Rome, and much of ancient Judaism and early Christianity. It is, indeed, the keynote of the 'modern' world, but it would be intolerably parochial of us to suppose that we invented it.

The later Heidegger, according to Lacoste, paints a very different picture. We now find ourselves, not wanderers in the world, but dwellers in the *earth*, inhabitants of some particular space, some territory (notice *'terra'* there) whose woods, and springs, and hilltops, are our shrines. We are country-dwellers, peasants, pagans, once again, in tune to voices, forces, vibrancies, to which we owe allegiance and of which we form a part. From Gaia to

Glastonbury and Greenpeace, we still inhabit Mother Earth. As with the 'subject–object' version, 'world' and 'earth', secularity and paganism, define each other, demand each other, sustain each other in existence.[48]

If human identity were simply a natural given, a product of its constituents, we might be simply stuck with this dialectic, doomed to oscillate between its poles. But, as I said at the beginning, human nature is as much a project as it is a given fact. Human identity includes the possibility of entertaining possibilities, of taking time together to work out how things are and how they might be made to be.

'Art', said Ernst Bloch, 'is a laboratory and also a feast of implemented possibilities.'[49] He had in mind Goethe's comment on Diderot's 'Essay on painting': 'the artist, grateful to nature, which also produced him, gives her a second nature in return, but one that is felt and thought and humanly perfected'.[50] Does not art, thus characterized, embrace science and technology, politics and ethics—all the things we do against 'the Background'? And is it not desirable and necessary, integral to *human* 'nature', that we should acknowledge this from time to time and, in so doing, give to contingency the form of celebration? Nor should we exclude the possibility that such celebration may be in praise of God.

Notes and references

1. John Searle, *The rediscovery of the mind* (MIT Press, London, 1992), p. 248.
2. Searle, *Rediscovery*, p. 127.
3. Some months after the conference at which this paper was first given, I found myself sitting next to Lord Longford at dinner, and asked him to verify the story. Memory undimmed at ninety, he did so with enthusiasm but (as so often happens with historical interpretation) the 'raw' version fits somewhat less neatly to my argument than the 'cooked'. 'So-and-so' was Sir Henry Maine, from whom Baldwin had learned, and had never forgotten, that 'the story of society is a story of evolution from contract to status. Or was it the other way round?'
4. G. W. H. Lampe, *God as spirit* (Clarendon Press, oxford, 1977), p. 226.
5. I regret not having yet had an opportunity to read John Searle, *The construction of social reality* (Penguin Books, London, 1995), but am encouraged by the extent to which my remark seems to chime in with his conclusion (see pp. 226–7).
6. Felix Wilfred, *From the dusty soil* (University of Madras, Madras, 1995), p. 71.
7. Searle, *Rediscovery*, p. 1; Gerald Edelman, *Bright air, brilliant fire: on the matter of the mind* (Penguin Books, London, 1992); the title of Edelman's second chapter is 'Putting the mind back into nature'.
8. Searle, *Rediscovery*, p. 3.
9. Searle, *Rediscovery*, p. 4 .
10. Anthony Kenny, *The metaphysics of mind* (Clarendon Press, Oxford, 1989), p. 7.
11. See Searle, *Rediscovery*, pp. 90–1.

12. Emile Durkheim, *Suicide: a study in sociology*, trans. John A. Spaulding and George Simpson (Routledge and Kegan Paul, London, 1952), p. 312.

13. W. J. Hankey, *God in himself: Aquinas' doctrine of God as expounded in the Summa theologiae* (Oxford University Press, Oxford, 1987), pp. 74–0.

14. See Fergus Kerr, 'Aquinas after Marion', *New Blackfriars*, **76**, 354–64 (1995). It would be difficult to exaggerate the importance of the fact that Aquinas's entire discussion, in the first part of the *Summa theologiae*, of God's simplicity, perfection, goodness, immutability, eternity, and even oneness, is announced as a discussion of 'the ways in which God does not exist' (Prologue to *S.t.*, Ia, q.3).

15. More generally, as David Burrell put it in his study of Ibn-Sina, Maimonides, and Aquinas: 'The unity of God can hardly be apprehended as a purely philosophical assertion. For the phrase itself—"God is one"—is but shorthand for specific confessions of faith, and our manner of elucidating it should reflect the shape of those confessions' (David B. Burrell, *Knowing the unknowable God: Ibn-Sina, Maimonides, Aquinas* (University of Notre Dame Press, Notre Dame, 1986), p. 111).

16. See Thomas Aquinas, *Summa theologiae*, Ia, 14.4, s.c.; Bernard, *De consideratione*, PL 182, 799.

17. John 14: 8, 9.

18. Henry Chadwick, *Boethius. The consolations of music, logic, theology, and philosophy* (Clarendon Press, Oxford, 1981), p. 212.

19. John 5: 19.

20. Searle, *Rediscovery*, pp. 96–7.

21. Augustine, *The Trinity*, ed. Edmund Hill (New City Press, New York, 1991), Book ii, 3; *In Joann. Ev.* (CCSL, 36), xxi, 4.

22. Lampe, *God as spirit*, p. 226.

23. Searle, *Rediscovery*, p. 85.

24. G. E. Lessing, 'On the proof of the spirit and of power'. In *Lessing's theological writings*, selected and introduced by Henry Chadwick (Adam and Charles Black, London, 1956), p. 53.

25. Karl Barth, *Protestant theology in the nineteenth century: its background and history* (SCM Press, London, 1972), p. 253.

26. I have set out a simplified version of Diderot's *Prospectus*, and a corresponding version of Bacon's, constructed from the table of contents of his *First part of the Great Instauration: the dignity and advancement of learning, in nine books*, in *The physical and metaphysical works of Lord Bacon*, ed. Joseph Devey (Henry G. Bohn, London, 1864), as appendices to Nicholas Lash, 'Reason, fools and Rameau's nephew', *New Blackfriars*, **76**, 368–77 (1995). Whereas Diderot set all theology, 'natural' and 'revealed', in that part of philosophy known as the 'Science of God', Bacon included only 'natural theology', leaving what he called 'sacred or inspired theology' quite outside the scheme of things to be surveyed from 'the small vessel of human reason'.

27. Bacon, *Advancement of learning*, Book II, Ch. 1.

28. Amos Funkenstein, *Theology and the scientific imagination from the Middle Ages to the seventeenth century* (Princeton University Press, Princeton, 1986), p. 41. (Which links up with what John Searle was saying: cf. *Rediscovery*, p. 20.)

29. Bacon, *Advancement of learning*, Book II, Ch. 1.

30. See Lessing, 'On the proof of the spirit and of power', p. 55.

31. See Brian Wicker, *The story-shaped world: fiction and metaphysics: some variations on a theme* (Athlone Press, London, 1975).

32. See Nicholas Lash, 'Ideology, metaphor and analogy', *Theology on the way to Emmaus* (SCM Press, London, 1986), pp. 95–119; this essay was first published in Brian Hebblethwaite and Stewart Sutherland (ed.), *The philosophical frontiers of Christian theology* (Cambridge University Press, Cambridge, 1982), pp. 68–94.

33. Searle, *Rediscovery*, p. 20.

34. Searle, *Rediscovery*, p. 55.

35. Richard J. Bernstein, *Beyond objectivism and relativism: science, hermeneutics, and praxis* (Basil Blackwell, Oxford, 1983), p. 18.

36. Stanley Cavell, *The claim of reason: Wittgenstein, skepticism, morality, and tragedy* (Clarendon Press, Oxford, 1979), p. 470.

37. See, for example, Searle, *Rediscovery*, pp. 20–1.

38. See, for example, Mary Midgley, *Science as salvation: a modern myth and its meaning* (Routledge, London, 1992).

39. Blaise Pascal, *Pensées*, 3.206.

40 Searle, *Rediscovery*, p. 191.

41. Gordon E. Michalson, *Lessing's 'ugly ditch': a study of theology and history* (Pennsylvania State University Press, London, 1985), p. 31.

42. Ludwig Wittgenstein, *Notebooks 1914–16*, ed. G. E. M. Anscombe and G. H. von Wright (Blackwell, Oxford, 1961), p. 86; quoted from Fergus Kerr, 'Aquinas after Marion', *New Blackfriars*, **76** (July/August), 364 (1995).

43. M. O'C. Drury, 'Some notes on conversations with Wittgenstein', in *Recollections of Wittgenstein*, ed. Rush Rhees (Oxford University Press, Oxford, 1984), p. 94; see pp. 92, 93.

44. Searle, *Rediscovery*, p. 175. He goes on to say that 'The work of the later Wittgenstein is in large part about the Background' (p. 177).

45. Searle, *Rediscovery*, p. 177.

46. Jean-Yves Lacoste, 'En marge du monde et de la terre: l'aise', *Revue de Métaphysique et de Morale* (1995), pp. 185–200.

47. Ephesians 2: 12.

48. See Lacoste, 'En marge du monde', p. 194.

49. Ernst Bloch, *The principle of hope*, trans. Neville Plaice, Stephen Plaice, and Paul Knight (Basil Blackwell, Oxford, 1986), Vol. 1, p. 216.

50. Quoted by Bloch, loc. cit.

CHAPTER 11

Binding the mind

PETER LIPTON

Several of the essays in this collection discuss the 'binding problem', the problem of explaining in neurophysiological terms how it is that we see the various perceptual qualities of a physical object, such as its shape, colour, location, and motion, as features of a single object. The perceived object seems to us a unitary thing, but its sensory properties are diverse and turn out to be processed in different areas of the brain. How then does the brain manage the integration?

Readers of the essays in this collection may find themselves suffering from an analogous binding problem about the study of consciousness, though this problem is conceptual rather than perceptual, and here the difficulty is to achieve the integration rather than to understand how an effortless integration is achieved. Consciousness is the ideal topic for interdisciplinary investigation. It is a central concern of such diverse disciplines as neurophysiology, evolutionary biology, psychology, cognitive science, philosophy, and theology, among others, yet none of these disciplines has come close to providing full answers to the central questions that consciousness raises. Interdisciplinary investigation seems an obvious way forward, but it generates the conceptual binding problem that this collection displays. The standard of the essays is high, but it is extraordinarily difficult to integrate their content into anything like a single picture. We are all apparently talking about the same phenomenon—the conscious awareness of the world that each of us enjoys first-hand—but it is quite unclear how to see the different things we say about this phenomenon as part of a single picture, or even as parts of different but compatible pictures.

Having raised the binding problem for the interdisciplinary study of consciousness, I hasten to say that I will not attempt even a partial substantive solution here: that is left as an exercise for the readers of this book. What I would like to do, however, is to suggest some ways that we may make progress on the problem by seeing the sort of structure its correct solution might have. Since my own specialty is the philosophy of science, this essay will focus on the relation between philosophical and scientific approaches to consciousness. One moral will be that part of the binding problem may be solved through unbinding: there should be a division of labour between scientists and philosophers, where scientists focus on the causal explanations of consciousness, leaving the recalcitrant problem of what consciousness itself is to the philosophers, who have had more practice at bumping their heads against brick walls. Another moral will be that apparently incompatible explanations of consciousness offered by different disciplines are sometimes compatible after all. The illusion of incompatibility arises because what appear to be different and incompatible answers to the same question are really different but compatible answers to different questions. I will illustrate one way this situation arises by describing some philosophical work on the structure of scientific explanations, work that brings out the importance of the contrastive structure of many why-questions.

Explaining correlations

We can begin with something that might be accepted by every contributor to this volume: there are correlations between types of brain activity and types or aspects of conscious states. Neurological research has uncovered some of these correlations, and there are presumably many more to come. A correlation, however, can be explained in many different ways. Here are four possibilities. First, it could be a mere coincidence, though the likelihood of this will fall as the known extent and frequency of correlation increases. I will assume that many mind–brain correlations are not mere coincidences. Second, the correlated states may be effects of a common cause. Thus the sound of thunder is correlated with the flash of light, and this is no coincidence. The sound does not, however, cause the light; rather they are both effects of the electrical discharge in the clouds. Some correlations between brain states and conscious states are of this sort, but again I assume that not all of them are. The third possibility is genuine causation, where one of the correlated states causes the other. This is illustrated by the correlation between the electrical discharge and the sound, and also by the correlation between the electrical discharge and the light. The final explanation of correlation I will consider is identity, where the states are correlated because they are in fact one in the same state. There is an excellent correlation between the presence of photons and the present of light, between molecular motion and temperature, between H_2O and water, and so on. The explanations of these correlations is not that photons cause light, that molecular motion causes temperature, or that H_2O causes water. They are rather that light just is a stream of photons, temperature just is molecular motion, and water just is H_2O. My observations about the division of intellectual labour on the problems of consciousness will focus on the differences between explaining mind–brain correlations in terms of causation and explaining them in terms of identity.

If we aim to explain consciousness, which sort of explanation should we seek? Should we aim for a causal explanation or an identity explanation, an account of what brings about conscious states or an account of what those conscious states themselves are? The traditional mind–body problem focuses on the identity question: are conscious mental states themselves physical or not? The physicalist claims that conscious states just are physical states; the dualist claims that they are not, however tight their causal links to the physical may be. Whatever answer we give to the identity question, however, the causal question remains. Whether mental states are physical or not they have, I take it, physical causes, and the question is what those causes are and how they operate. So what we really want are both sorts of explanation of consciousness. This is, however, where I suggest a division of labour. Scientists should tackle the causal question, leaving the identity question to torment the philosophers.

My reason for warning scientists off attempting identity explanations of mind–brain correlations is not that scientists are in general ill-equipped to discern identities. On the contrary, scientists are the pre-eminent dis-coverers of deep and informative identities, as my examples above of light, temperature, and water illustrate. Moreover, both scientists and laypeople are in general surprisingly good at determining whether a particular correlation is due to coincidence, common cause, causation, or identity. Philosophers and psychologists are rather less good at saying just how we manage this cognitive feat, but some of our techniques are relatively clear. For example, if we want to determine whether a correlation obtains because of causation or because there is a common cause, we may manipulate the one type of correlated state to see whether there is a reaction in the other. Barometer readings may be well correlated with subsequent weather, but unfortunately we find that artificially changing the readings does not change the weather, so barometers don't cause the weather. We also have some relatively straightforward techniques for distinguishing causal from identity correlations. For example, if we find that correlated states have different locations, we know the correlation cannot be a matter of identity, since 'two' states can be identical only if they share all their properties. Causally connected states may be separated in space and time; identical states cannot be. Nevertheless, conscious mental states have some peculiar features that make the empirical determination of identity claims especially difficult. Many of these features are discussed in other essays in the volume, but it is perhaps useful to bring some of them together here. My aim is not to show that the physicalist's identification of conscious and physical states is mistaken, but just that the issue is contentious and strangely resistant to the normal empirical techniques that scientists deploy.

Identity problems

There are several ways of bringing out the difficulties of a physical identification. One is to point to features of consciousness that no purely physical state could apparently possess. It is tempting to say here that there is one core feature that meets this description, and it is simply the consciousness of consciousness, the fact that experiences are, well, experiences. How could this experience I am now enjoying literally be a neural state? The problem, for those who take it to be a problem, is not just that the identity claim seems false, but that we do not understand what it would be for it to be true, which is to say that we do not understand what the identity claim means. This situation is quite different from that of the standard scientific identities mentioned above. You may be amazed to be told that temperature is molecular motion, you may even deny it, but you understand what is being claimed. The mind–brain identity claim, by contrast, does not seem even to make sense.

Or so say some. Other thoughtful people find no difficulty here, and accuse those who claim that the very state of being conscious could not be a feature of any purely physical system of begging the question. Why should consciousness not just be, say, the synchronous firing of certain neurones at a certain rate? In my view the claimed difficulty in understanding the identity claim does not in fact beg the question, but it obviously will not move anyone who does not feel it. So those who are troubled by the identity claim have looked for other more specific and more clearly recalcitrant features of conscious states. Some have pointed to the fact that many conscious states have content: they are representational, they are about something. Perceptual states are a good example. When someone sees a chair, we can distinguish between the perceptual states itself and what that state is about, its representational content. Strangely, perhaps, this contrast can be made even if the perception is an hallucination. Even here, the perception points beyond itself to a chair, which, in the case of hallucination, just happens not to exist. Physical states, it is claimed, cannot have this feature. A piece of chalk, for example, has many properties, but it is not about anything, it does not represent, something only mental states can do. To this the physicalists may respond by pointing to the marks on the board that the chalk can be used to produce. Those marks are purely physical, yet being sentences, serve to represent. Then the dualist may reply that the physical representation is real but inevitably parasitic on the originating representational powers of the minds that perceive the marks. And so the debate continues.

A second claimed peculiarity of conscious states is that they are subjective and perspectival. Unlike physical states, conscious states can only exist in virtue of being experienced, and their existence thus incorporates a certain point of view, the point of view of the creature having the experience. A third and related peculiarity is the so-called asymmetrical access that we have to conscious states, again apparently unlike any physical states. The physical world includes things like chairs that are publicly observable, and things like electrons that are unobservable, and everything physical is either one or the other. Conscious states seem to be neither. My conscious states seem to be immediately accessible to me, but to no one else. To other people, my conscious states seem rather like electrons, whose existence can only be inferred, whereas I seem to have a more direct and quite different sort of access to my own consciousness, as you do to yours.

All these features of consciousness—its very existence, its representational content, its subjective and perspectival nature, and our asymmetrical access to it—have been claimed to stand in the way of any identification of conscious and physical states. Another way of bringing out the difficulty of identification is to stress the apparent contingency of any correlation between mental and physical states. No matter how much physical information is provided, it never seems necessary that consciousness should come along with it. Similarly, whatever behavioural or evolutionary story one describes, it never seems necessary that it be accompanied by consciousness. It always seems possible that the same behaviour or the same evolutionary history might have existed without any accompanying or resulting experience. The problem is completely general: we are given a sophisticated biological story, and then it is claimed that this or that behaviour could be performed only if the creature were conscious. That is the part that never seems convincing: it always seems possible that the behaviour (or the neural activity, or the evolutionary history) could exist without the existence of the conscious states. This persistent contingency stands in the way of seeing how we can identify mental and physical states, and because the feeling of contingency remains not just for the physical facts we now know, but for any physical facts we can imagine, the resolution of the difficulty does not appear to be one that further scientific research could resolve.

Keeping to causation

The moral I draw from the formidable difficulties in making out a mind–brain identity claim is not that physicalism is false: perhaps they can be overcome.

In any event, the dualistic alternatives to physicalism, which hold that mental states have an irreducibly non-physical component, face many difficulties of their own. The moral is rather that scientists should leave the identity question to the philosophers. It is not the sort of question that empirical inquiry is suited to solve, and it is precisely the sort of question that is the philosophers' business. This is not, of course, to say that philosophers will ever provide a definitive answer, but philosophers, unlike scientists, make their business out of the insoluble.

Many scientists are aware of the difficulties I have flagged concerning the identity of consciousness and are only too happy to leave them to the philosophers, but some over-react, taking the view that consciousness is not a fit subject for science. This is a mistake. Scientists should study consciousness, but they should seek causal explanations, not identity explanations. For the difficulties disappear or are at least substantially diminished if we shift our attention from the question of what mental states are to questions of their aetiology.

Why should the situation improve so dramatically if we ask causal questions? The identity difficulties arise, as we have seen, because conscious states seem to have features that no purely physical system could possess, and because any connection between physical and conscious states appears to be contingent, unlike the necessity that the identity of a thing with itself requires. Neither of these points is a barrier to causal connection. There is no presumption that causes resemble their effects. Pushing buttons and flicking switches cause an enormous variety of effects, with properties not to be found in the switch or button itself. Moreover, the fact that a cause might not have produced its effect provides no reason to say that it was not really a cause. The perennial possibility of interference and breakdown make this clear, as switches and buttons again illustrate. Perhaps there is no behaviour we perform which we could not imagine being performed without consciousness, but we know that we do these things with conscious states, and it is up to the scientists to tell us what in our brains causes those states to occur. By sticking to the causal questions, scientists will also avoid making identity claims that appear to explain consciousness only by explaining it away, denying its essential experiential component, and avoid the sort of implausible 'nothing but' reductionism that is criticized in some of the essays in this book.

John Searle is another champion of the scientific study of the causes of consciousness, as his essay shows (Chapter 2). There is, however, an important difference between us. Searle holds that causes and effects need not be distinct, and that scientists ought to look for causes of mental states that also tell us what those states are, causes that answer the identity question. Thus he must reject my proposed division of labour, where the scientists leave the identity question to the philosophers. According to Searle, scientists ought to be answering both questions. So it may be useful for me to say just where I disagree with him and why.

Searle makes a helpful distinction between two versions of the subjective/objective distinction—an epistemic version and an ontological version—and uses that distinction to reveal a fallacy in a common argument against the possibility of scientific study of consciousness. The fallacious argument is that science is objective, consciousness is subjective, so science can't study consciousness. The source of the fallacy is an equivocation between the two versions of the subjective/objective distinction. Science is objective in the epistemic sense: its investigations lead to intersubjective agreement about objective facts. Consciousness is subjective, in the ontological sense, which is to say that consciousness states must be experienced to exist. Ontological subjectivity does not, however, entail epistemic subjectivity, so the conclusion of the argument doesn't follow. It remains possible to have an epistemically objective, scientific investigation into the nature of ontologically subjective conscious states.

I agree with Searle that the argument against the possibility of the scientific study of consciousness is fallacious, and for the reason he gives. I also agree about the nature of the subjectivity of consciousness: it is ontological, not epistemic. But it is this ontological subjectivity and its consequences that are at the basis of the identity difficulties I have mentioned. Searle, by contrast, holds that these are only difficulties in the context of one model of identities: there is another model for which they do not arise. He illustrated the first model with the example of the identity of heat and molecular motion, the second with the identity of solidity and the vibratory movement of molecules in lattice structures. I disagree with Searle here, because I think that the difficulties arise for both models.

Searle claims, rightly in my view, that the identity of heat and molecular motion is a bad model for consciousness, because the truth of that identity depends on conceiving of heat in a way that disassociates it from the feeling of heat. (This point has also been emphasized by Saul Kripke, in Lecture III of his *Naming and Necessity* (1980).) What makes this identity unproblematic is that we sharply distinguish heat or temperature from the conscious sensation of heat. It is not the sensation that is claimed to be identical to molecular motion. (For all I know, the molecules in my brain move more rapidly when I touch something very cold than when I touch something whose temperature is close to that of my own body.) The identity quite correctly treats heat or temperature as something out there, stripped of the feeling, and this makes possible the strict identification with the physical state. Many other scientific identity claims work in the same way. This strategy of stripping off the feeling is, however, obviously hopeless if it is the feeling that we are trying to identify, which is exactly what we are trying to do if we attempt a physical identification of consciousness. This is then our dilemma: if we strip off the feeling, we will lose the very thing we are supposed to be identifying; if we leave the feeling in, we cannot see how the identity could hold.

Searle, however, rejects the second horn of this dilemma, suggesting that we can understand a proper mind–brain identity if we use a different analogy, such as the identity between solidity and vibration within a lattice. This is where we part company. The point that Searle and others have made about heat and molecular motion seems to me to apply equally to cases such as solidity. It may be that solidity is in some important ways a different case from heat. Perhaps solidity is a more 'holistic' property than heat; for example, it makes no sense to say that a single molecule is solid. On the other hand, it may not make sense to say that a single atom has a particular temperature either; perhaps only an ensemble of atoms can have a temperature. An interesting difference between solidity and heat may be that solidity depends on physical structure (the lattice) in a way that heat does not. In any case, the salient similarity remains. Just as in the case of heat and molecular motion, we understand the identification of solidity with vibration within a lattice only by shaving off the feeling, in this case the feeling of solidity or resistance to pressure. The physical identity works by shaving off the feeling, whether it is the sensation of heat, or the sensation of resistance that solid objects afford. So the case of solidity does not help us to see how a physical identification of conscious states is possible. Nor would it help to chose another analogy, such as the identity between light outside the visible range and photons with certain specified energies, where there is no corresponding sensation to strip. This brings us no closer to seeing how the peculiarities of consciousness could be tamed with a physical identification. This is the reason I continue to see a deep difference between the causal and identity questions, and support the division of intellectual labour.

I have argued that the metaphysical difficulties of the philosophers' mind–body problem are no bar to the scientific explanation of consciousness, where the explanations are causal. Some scientists have, however, held that such explanations are not to be had, for epistemological reasons. These are in my view bad reasons, and I now want to suggest why. One such reason is based on the equivocation over the claim that consciousness is a subjective phenomenon that Searle effectively exposes. Another concerns a feigned modesty about the scope of scientific inquiry. To see what this comes to, we may usefully return to the topic of correlations. As I have already observed, scientists are in general at least as sensitive as the rest of us to the distinction between the different explanations for correlations, such as coincidence, causation, and identity. Nevertheless, in their philosophical moments, some scientists claim that all science can ever deliver are the correlations themselves, a position that would leave genuine causal explanations of consciousness outside their purview.

Why do they say this? Part of the explanation may be the legacy of logical positivism, a philosophical movement from the early part of this century with clear antecedents in the earlier history of philosophy. Positivism has attracted many philosophically inclined scientists, partially because, unlike most philosophical positions, it appeared to offer some useful methodological advice about scientific research, and perhaps also partially because it was a philosophy that held up science as the acme of human endeavour. Paradoxically, however, positivism has also lead some scientists drastically to understate what science can achieve.

The positivists maintained that only claims that have empirical consequences are really claims at all; they are the only sentences that say anything. If a sentence has no empirical consequences, it is just noise: it can not even rise to the level of being false. This is already an implausible view, but the positivists went on to say something much more implausible. They said that the meaning of a statement that does have empirical consequences just is those empirical consequences. One of the slogans of the time was, 'The meaning of a statement is its method of verification'. This is a serious mistake, a confusion of the claim one is making with the evidence one might have for that claim. This confusion arises in many different areas in the history of science in this century, and certainly in psychology. As Mary Midgley points out in her essay (Chapter 8), behaviourism is a particularly pure example of the confusion. The positivist view that all that a claim can describe is the evidence, combined with the view that the only evidence for other people's mental states is their behaviour, yields behaviourism. This is the position that talk about mental states is really just talk about behaviour. This position, along with one of its major weaknesses, is summed up in the familiar story about the two behaviourists who meet in the street. One says to the other, 'You're fine. How am I?'

Behaviourism is not in fashion in psychological circles these days, but it is not a dead horse. The positivistic impulse remains among many philosophers and scientists, though it tends to find less obvious expression than it did in the case of behaviourism. It is the impulse to confuse the evidence with what it is evidence for. Only what is in some sense observable can be scientific evidence, but scientists can have evidence for what is unobservable. Here for once physics provides a good model. What physics shows us is that we can have good evidence for claims that go way beyond not only that we in fact observe but what is even in principle observable. Conscious states may not be intersubjectively observable, and perhaps there is even an interesting sense in which causal relations are in general unobservable, but these are not good reasons to deny that science can study the causes of conscious states, any more than the unobservability of atomic interactions provides a reason to deny that physicists may profitably study them.

Explanatory pluralism

Having made my pitch for the causal explanation of consciousness, I want to conclude with a few observations about the nature and structure of causal explanations. The essays in this book make a number of important contributions to this subject. Many of them bring out what we might call the pluralism of causal explanation. Behind every event, and so behind every conscious state, there is a long and dense causal history. No explanation can capture all of it, and no explanation needs to. But not just any cause will serve to answer a particular question about a given event. The Big Bang is part of the causal history of every event, but does not explain all of them. So we need some sense of what conditions a cause has to satisfy to answer a particular question. That is the problem of causal selection. A sensitivity to the plurality of causal explanations also helps us to see that different explanations of consciousness, which may seem competitors, are sometimes really compatible: they are just answering different questions.

The contributors to this book develop themes of causal selection and explanatory pluralism in various ways. Margaret Boden (Chapter 1) brings out the contrastive 'rather than' structure of many causal explanations, arguing that adequate explanations of conscious states must specify causes that explain why the state takes one form rather than another. One of the many aspects of explanation that Stephen Rose discusses (Chapter 5) concerns the way the plurality of explanation stratifies into different levels of explanation, and the importance of finding the level appropriate to conscious states and the questions we ask about them. The idea of the multiplicity of levels of explanation is also developed by Mary Midgley (Chapter 8) and by Fraser Watts (Chapter 9) both of whom stress the importance of offering explanations at different levels and the difficulty of seeing how these explanations could be integrated into a single picture of the aetiology of consciousness.

All of these observations raise important issues for the explanation of consciousness, but here I will just say a bit more about one of them, the importance of the contrastive structure of causal explanations. As I have already mentioned, different explanations of consciousness sometimes give the illusion of incompatibility when they are really compatible but answering different questions. One of the reasons two such explanations may seem incompatible is that they are both answers to questions that could be phrased as 'Why X?' for the same X, and so seem to be answering the same question. Yet they are answering different questions. The contrastive structure of explanation shows how this is possible. In many cases, what we are really asking is not just, 'Why X?', but rather, 'Why X rather than Y?', and this may be a different question, requiring a different answer, than the question, 'Why X rather than Z?'. This is another one of those deep truths about explanation that we learned as children: 'Why do birds fly south in the winter? Because it is too far to walk.' It spoils the joke to explain it, but what this and many other children's jokes depend on is an unexpected contrast switch. The intended question was, 'Why do birds fly south in the winter rather than stay where they are?', whereas the reply answers the different question, 'Why do birds fly south in the winter rather than walk?' Both questions cite the same fact—birds fly south in the winter—but they use different foils, requiring different answers. The two answers are compatible, but they answer different questions. In this example the joke works because the difference in contrast is so obvious, but for many of the more intellectually demanding 'why' questions we ask this is not so. We focus only on the fact, and then reject someone else's answer because it doesn't answer our real question, without realizing that it may nonetheless be a legitimate answer for someone asking a different contrastive question about the same fact. We can see these different answers as compatible, once we appreciate the aspect of the plurality of causal explanations that the relativity of explanation to contrast reveals.

In addition to showing how apparently incompatible explanations can sometimes be compatible, an appreciation of the way contrastive explanation works shows how even a very partial causal story can sometimes provide an adequate explanation. The case of syphilis and paresis illustrates the point. Paresis can be caused only by tertiary untreated syphilis, but most people with tertiary untreated syphilis fortunately do not go on to contract paresis. This causal situation has provided the scene for a raging dispute in the philosophical literature on explanation. The question is whether the fact that a person has paresis can be explained by pointing out that he previously had tertiary untreated syphilis. Some philosophers say yes, on the grounds that the syphilis is a cause of the paresis and causes explain effects. Other philosophers say no, on the grounds that the paresis could not be deduced, or even shown to be probable, on the basis of the syphilis, since most people with syphilis do not go on to get paresis. The syphilis is a cause, but it is too small a part of the total cause to be explanatory. The dispute can be settled if we bring in contrastive questions. If the question is, 'Why did Smith rather than Jones get paresis?', and Jones does not have syphilis, then Smith's syphilis provides a good answer. If, on the other hand, the question is, 'Why did Smith rather than Doe get paresis?', where both of them had syphilis, Smith's syphilis is not explanatory. Both questions cite the same fact but a different foil, and the cause that answers one of the questions will not answer the other. This again illustrates the sensitivity of explanation to contrast, and shows how even a very partial cause can be explanatory. In many cases, all it takes to explain a contrasts is a cause that makes a difference between the fact and the foil.

The contrastive structure of explanations thus reveals that causal explanation is sometimes easier than it may at first appear. One benefit of seeing this in the context of the explanation of consciousness is that it helps further to alleviate the worry about the apparent contingency of the link between mental and physical states. I have already suggested that this worry is reduced if we switch from identity to causal explanations. It is further reduced by an awareness of the way explanations are sensitive to contrasts. Perhaps none of the scientific accounts now available really does explain why an organism has a particular conscious state rather than being an unconscious robot. Even if this is so, however, they may explain why the organism has that particular conscious state rather than another one. The second question may be easier to answer, because it presupposes what the first question asks about, namely that the organism is conscious at all. An awareness of the contrastive structure of explanation should also encourage us to look more carefully at our questions before we assess the answers, and in particular to take care not to ask one contrastive question and then proceed to answer another, lest we inadvertently offer the grown-up equivalent of the joke about the birds.

My discussion of contrastive explanation takes me back to the conceptual binding problem with which this essay began, the problem of integrating the diverse interdisciplinary approaches to consciousness. The request for integration has two parts and I have responded to them differently. The first part is a request to be shown how the different answers to questions about consciousness are compatible. Of course some of them just are not compatible, but I have suggested that there is more widespread compatibility that may at first appear, and that we can see this by distinguishing different types of explanation. We need to distinguish causal explanations from identity explanations, and we need to us the contrastive structure of explanation to distinguish the many different causal questions that may all be questions about the same facts. By making these distinctions, we will find compatibility in some places where there initially seemed to be competition. The second part of the request for integration asks for more: it asks that we show not just that different answers are compatible, but that we fit them all together into a single, unified account of the mind. This is the request that I have resisted.

At our present stage of understanding, what is needed is division of labour, not superficial attempts at a global picture. The disciplines concerned with consciousness have plenty to learn from each other, as the essays in this book show, but in my judgement the way forward for the foreseeable future is through the simultaneous development of diverse approaches at different levels, rather than through the attempt at a single, big picture.

Index

OXFORD
1 JAN 2000
UNIVERSITY